Transport and Chemical Transformation
of Pollutants in the Troposphere

Series editors: Peter Borrell, Patricia M. Borrell, Tomislav Cvitaš,
Kerry Kelly and Wolfgang Seiler

Springer

Berlin
Heidelberg
New York
Barcelona
Budapest
Hong Kong
London
Milan
Paris
Santa Clara
Singapore
Tokyo

Transport and Chemical Transformation
of Pollutants in the Troposphere

Volume 10

Photo-oxidants, Acidification and Tools: Policy Applications of EUROTRAC Results

The Report of the EUROTRAC Application Project

Peter Borrell, Peter Builtjes, Peringe Grennfelt, Øystein Hov
Editors and Coordinators

Members

Roel van Aalst, David Fowler, Gérard Mégie,
Nicolas Moussiopoulos, Peter Warneck, Andreas Volz-Thomas,
Richard P. Wayne

Springer

Editors

Dr. PETER BORRELL
EUROTRAC ISS
Fraunhofer Insitute
Kreuzeckbahnstr. 19
D-82467 Garmisch-Partenkirchen

Prof. PETER J. H. BUILTJES
TNO-MEP
P.O. Box 60 11
NL-2600 JA Delft

Dr. PERINGE GRENNFELT
Swedish Environmental Research Institute
Box 47086
S-40258 Göteborg

Prof. ØYSTEIN HOV
Norwegian Institute for Air Research
P.O. Box 100
N-2007 Kjeller

With 24 Figures and 18 Tables

ISBN 3-540-61783-3 Springer-Verlag Berlin Heidelberg New York

Library of Congress Cataloging-in-Publication Data. Photo-oxidants, acidification and tools: policy applications of EUROTRAC result: the report of the EUROTRAC application project / Peter Borrell ... [et al.]. p. cm. – (Transport and chemical transformation of pollutants in the troposphere; v. 10). Includes bibliographical references and index. ISBN 3-540-61783-3 (hc) 1. Tropospheric chemistry. 2. Photochemical oxidants – Environmental aspects – Europe. 3. Acidification – Environmental aspects – Europe. 4. Environmental policy – Europe. 5. Environmental protection – Europe. I. Borrell, Peter. II. Series. QC881.2.T75P46 1996 363.738'7 – dc21 96-45984 CIP

Cover Design: Struve & Partner, Heidelberg

Coverpicture from Fotoverlag Huber, D-82467 Garmisch-Partenkirchen

SPIN 10514726   30/3136-5 4 3 2 1 0 – Printed on acid-free paper

Transport and Chemical Transformation of Pollutants in the Troposphere

Series editors: Peter Borrell, Patricia M. Borrell, Tomislav Cvitaš, Kerry Kelly and Wolfgang Seiler

Foreword by the Series Editors

EUROTRAC is the European co-ordinated research project, within the EUREKA initiative, studying the transport and chemical transformation of pollutants in the troposphere. The project has achieved a remarkable scientific success since its start in 1988, contributing substantially both to the scientific progress in this field and to the improvement of the scientific basis for environmental management in Europe. EUROTRAC, which at its peak comprised some 250 research groups organised into 14 subprojects, brought together international groups of scientists to work on problems directly related to the transport and chemical transformation of trace substances in the troposphere. In doing so, it helped to harness the resources of the participating countries to gain a better understanding of the trans-boundary, interdisciplinary environmental problems which beset us in Europe.

The scientific results of EUROTRAC are summarised in this report which consists of ten volumes.

Volume 1 provides a general overview of the scientific results, prepared by the Scientific Steering Committee (SSC) and the International Scientific Secretariat (ISS) of EUROTRAC, together with brief summaries of the work of the fourteen individual subprojects prepared by the respective subproject coordinators.

Volumes 2 to 9 comprise detailed overviews of the subproject achievements, each prepared by the respective subproject coordinator and steering group, together with summaries of the work of the participating research groups prepared by the principal investigators. Each volume also includes a full list of the scientific publications from the subproject.

The final volume, 10, is the complete report of the Application Project, which was set up in 1993 to assimilate the scientific results from EUROTRAC and present them in a condensed form so that they are suitable for use by those responsible for environmental planning and management in Europe. It illustrates how a scientific project such as EUROTRAC can contribute practically to providing the scientific consensus necessary for the development of a coherent atmospheric environmental policy for Europe.

A multi-volume work such as this has many contributors and we, as general editors, would like to express our thanks to all of them: to the subproject coordinators who have borne the brunt of the scientific co-ordination and who have contributed so much to the success of the project and the quality of this report; to the principal investigators who have carried out so much high-quality scientific work; to the members of the International Executive Committee (IEC) and the SSC for their enthusiastic encouragement and support of EUROTRAC; to the participating governments in EUROTRAC, and in particular the German Government (BMBF) for funding, not only the research, but also the ISS publication activities; and finally to Mr. Christian Witschell and his colleagues at Springer Verlag for providing the opportunity to publish the results in a way which will bring them to the notice of a large audience.

Peter Borrell *(Scientific Secretary, ISS)*   EUROTRAC ISS
Patricia May Borrell   Fraunhofer Institute (IFU)
Tomislav Cvitaš   Garmisch-Partenkirchen
Kerry Kelly
Wolfgang Seiler *(Director, ISS)*

Table of Contents

EUROTRAC Application Project

Participating Scientists

Dr. Peter Borrell (*Coordinator*)
IFU, Garmisch-Partenkirchen

Photo-oxidant Group

Professor Øystein Hov *(Convener)*
University of Bergen

Dr. Roel M. van Aalst
RIVM, Bilthoven

Professor Nicolas Moussiopoulos
Aristotle University, Thessaloniki

Dr. Andreas Volz-Thomas
KFA, Jülich

Acidification and Nutrification Group

Dr. Peringe Grennfelt *(Convener)*
Swedish Environmental Research Institute, Göteborg

Dr. David Fowler
Institute of Terrestrial Ecology, Edinburgh

Professor Peter Warneck
Max-Planck-Institut für Chemie, Mainz

Tools Group

Professor Peter J. H. Builtjes *(Convener)*
IMW-TNO, Delft

Dr. Gérard Mégie
CNRS, Paris

Dr. Richard P. Wayne
University of Oxford

Authors names and addresses

Dr. Roel M. van Aalst
RIVM
(EEA Topic Centre: Air Quality)
P.O.Box 1
NL-3720 BA Bilthoven
Netherlands

Dr. Peter Borrell
EUROTRAC ISS
Fraunhofer-Institut (IFU)
Kreuzeckbahnstrasse 19
D-82467 Garmisch-Partenkirchen
Gremany

Prof. Peter J. H. Builtjes
TNO-MEP
P.O. Box 6011
NL-2600 JA Delft
Netherlands

Dr. David Fowler
Institute of Terrestrial Ecology
Bush Estate, Penicuik
GB-Midlothian EH32 ORA
United Kingdom

Dr. Peringe Grennfelt
Swedish Environmental Research
   Institute (IVL)
Box 47086
S-40258 Göteborg
Sweden

Prof. Oystein Hov
Norwegian Inst. for Air Research
(NILU)
Instutittveien 18
PO Box 100
N-2007 Kjeller
Norway

Prof. Nicolas Moussiopoulos
Aristotle University Thessaloniki
Box 483
GR-54006 Thessaloniki
Greece

Dr. Gérard Mégie
CNRS ,B.P. 3
F-91371 Verrières Le Buisson
France

Dr. Andreas Volz-Thomas
Forschungzentrum Jülich
Postfach 1913
D-52425 Jülich
Germany

Prof. Peter Warneck
Max-Planck-Institut für Chemie
Postfach 3060
D-55020 Mainz
Germany

Dr. Richard P. Wayne
Oxford University
Physical Chemistry Laboratory
South Parks Road
GB-Oxford OXl 3QZ
Great Britain

Acronyms and abbreviations

ABL	Atmospheric Boundary Layer
ACE	Acidity in Cloud Experiments (EUROTRAC subproject until 1992)
ACDEP	Acid deposition
ADOM	Acid Deposition and Oxidant Model
ALPTRAC	High Alpine Aerosol and Snow Chemistry Study (EUROTRAC subproject)
AP	Application Project (EUROTRAC project)
APSIS	Athenian Photochemical Smog Intercomparison of Simulations
ASAM	An Integrated Assessment Model for Acidification (Imperial College, London)
ASE	Air-Sea Exchange (EUROTRAC subproject)
ASG	Application Steering Group (EUROTRAC working group)
BAT	Best Available Technology
BIATEX	Biosphere-Atmosphere Exchange of Pollutants (EUROTRAC subproject)
CEC	Commission of the European Communities (now the EC)
CCN	Cloud Condensation Nuclei
CFC	Chlorofluorocarbon
CG	Cloud Group (EUROTRAC working group)
CIT	Carnegie Inst. of Technology Multiphase Transport Model
CMWG	Chemical Mechanism Working Group (EUROTRAC working group)
CORINE	Co-ordinated Information System on the State of the Environment and Natural Resources (An EU project now subsumed within the EEA)
CORINAIR	The atmospheric part of CORINE
CSAM	An Integrated Assessment Model for Acidification (Stockholm Environmental Institute)
CSU	Colorado State University
DIAL	Differential absorption lidar
DMS	Dimethyl sulphide
DMSO	Dimethyl sulphoxide
DMSP	Dimethylsulphopropionate
DOAS	Long path differential optical absorption spectroscopy
DoE	Department of the Environment (UK)

DRAIS	Dreidimensionales Regionales Ausbreitungs- und Immissions-Simulationsmodell
EC	European Commission
ECE	see UN-ECE
ECHAM	ECWMF Model (Hamburg)
ECN	Netherlands Energy Research Foundation (Petten)
ECMWF	European Centre for Medium Range Weather Forecasting
EEA	European Environmental Agency
EMEP	The co-operative programme for monitoring and evaluation of the long range transport of air pollutants in Europe; set up under the LRTAP.
EPA	Environmental Protection Agency (USA)
EU	European Union
EUMAC	European Modelling of Atmospheric Constituents (EUROTRAC subproject)
EURAD	The European Regional Acid Deposition Model; the main model used in EUMAC
EUREKA	A political initiative to encourage the trans-national development of technological research and development in Europe. (Founded in 1985)
EURORADM	The RADM model adapted for "European chemistry"
EUROTRAC	European Experiment on the Transport and Transformation of Environmentally Relevant Trace Constituents in the Troposphere over Europe. (A EUREKA environmental project)
EZM	EURAD Zooming Model
FIELDVOC	Field Studies of the Tropospheric Degradation Mechanisms of Biogenic VOCs.
FINOX	Finnish limited area model for oxidised nitrogen compounds
FOG	University of Munich Fog Model
GAW	Global Atmosphere Watch
GC	Gas Chromatography
GCE	Ground-based Cloud Experiments (EUROTRAC subproject)
GENEMIS	Generation of European Emission Data (EUROTRAC subproject)
GLOBE	RIVM report
GLOMAC	Global Modelling of Atmospheric Chemistry (EUROTRAC subproject)
GWP	Global Warming Potential
HALIPP	Heterogeneous and Liquid Phase Processes (EUROTRAC subproject)
HBFC	Hydrobromofluorocarbon
HCFC	Hydrochlorofluorocarbon

HELCOM	Helsinki Commission for the Protection of the Baltic Marine Environment
HIRLAM	High Resolution Limited Area Model
IEC	International Executive Committee (EUROTRAC)
IIASA	International Institute for Applied Systems Analysis, Laxemburg, Austria
IPCC	Intergovernmental Panel on Climate Change
ISS	International Scientific Secretariat (EUROTRAC)
ITE	Institute for Terrestrial Ecology (Edinburgh, UK)
JETDLAG	Joint European Development of Tunable Diode Laser Absorption Spectroscopy for the Measurement of Atmospheric Trace Gases (EUROTRAC subproject)
KAMM	Karlsruhe Meteorological Model
KFA	Environmental Research Institute (Jülich, Germany)
KNMI	Royal Netherlands Meteorological Institute
LACTOZ	Laboratory Studies of Chemistry Related to Tropospheric Ozone (EUROTRAC subproject)
LCC	Lurman, Carter and Coyner chemical mechanism
LOTOS	Long Term Ozone Simulation (A model for the study of photo-oxidants)
LRTAP	Geneva Convention on the Long-Range Transport of Air Pollution
MARS	Model for the Atmospheric Dispersion of reactive Species
MEMO	A Non-hydrostatic Mesoscale Model
MM4/MM5	The meteorological drivers used for the EURAD model
Moguntia	A global model extensively used in GLOMAC
MOZAIC	Measurement of Ozone by Airbus In-Service Aircraft
NCAR	National Centre for Atmospheric Research (Boulder, USA)
NDSC	Network for the Detection of Stratospheric Change
NMHC	Non-Methane Hydrocarbon
NO_x	the compounds NO and NO_2
NO_Y	NO_x and other nitrogen compounds formed from NO and NO_2
NO_z	NO_Y -NO_x
OCTA	An EC photo-oxidant project
ODP	Ozone Depletion Potential
OECD	Organisation for Economic Co-operation and Development
OXIDATE	Oxidant Data Collection in OECD Europe 1985-87
PAN	Peroxyacetylnitrate

PARCOM	Paris Commission for the Environmental Protection of the North East Atlantic
PHOXA	Photochemical oxidant and Acid Deposition
PI	Principal investigator within a EUROTRAC subproject
POCP	Photo-Oxidant Creation Potential
POLLUMET	Pollution and Meteorology
PORG	Photo-oxidant Review Group (UK DoE)
PTD	Photo-Thermal Deflection
RADM	Regional Acid Deposition Model; the US predecessor of EURAD and EURORADM developed at NCAR
RAINS	Regional Acidification and Simulation Model (developed at IIASA)
REKLIP	Regio-Klima-Projekt
REM-III or 3	see RTM-III
RIVM	National Institute of Public Health and Environmental Protection (Netherlands)
RTM-III	Regional Transport Model-3
SERCO	The UK company which conducted the review of EUROTRAC in 1991/2
SPONS	Swiss Plateau Ozone Simulation exercise
SSC	Scientific Steering Committee (EUROTRAC)
STREAM	Stratosphere and Troposphere Experiments by Aircraft Measurements
TADAP	Transport and Acid Deposition Model
TDLAS	Tunable-diode laser spectroscopy
TESLAS	Joint European Programme for the Tropospheric Environmental Studies by Laser Sounding (EUROTRAC subproject)
TM-2	Transport Model-2
TOASTE	Transport of Ozone and Tropospheric Exchange
TOPAS	Tropospheric Optical Absorption Spectroscopy (EUROTRAC subproject)
TOR	Tropospheric Ozone Research (EUROTRAC subproject)
TRACT	Transport of Pollutants over Complex Terrain (EUROTRAC subproject)
TRANSALP	A subsidiary project within TRACT
TREND	Dutch atmospheric and transport model used for trend analysis
TROLIX	Tropospheric Ozone Lidar Intercomparison Experiment (A TESLAS field campaign)
UAM	Urban Airshed Model

UK	United Kingdom
UN	United Nations
UN-ECE	United Nations Economic Commission for Europe
UNEP	United Nations Environmental Programme
US	United States
UV	Ultra violet
UV-B	Ultra violet-B radiation (ca 280 to 310 nm)
VOC	Volatile organic compound
WHO	World Health Organisation (UN)
WMO	World Meteorological Organisation (UN)
ZONS	Zürich Ozone Simulation exercise

Introduction

EUROTRAC is the EUREKA environmental project studying the transport and chemical transformation of pollutants in the troposphere over Europe. At its inception in 1988 it had three aims:

* to increase the basic understanding of atmospheric science;

* to promote the technological development of sensitive, specific, fast-response instrumentation for environmental research and monitoring; and

* to improve the scientific basis for taking future political decisions on environmental management within Europe.

It was clear at an early stage, as the fourteen subprojects were formed and more than two hundred research groups in twenty-four countries were incorporated, that the first aim would readily be achieved. An ample demonstration that the early indications were correct is provided in the other volumes in this series which describe the scientific progress made. Substantial progress was also made towards achieving the second aim although some problems were encountered, mainly due to the high cost of the technological development required.

However as the project progressed it was realised within EUROTRAC, and highlighted by an external review, that the third goal would not be achieved unless some special mechanism was devised to facilitate it. So in 1993 the International Executive Committee formed the Application Project (AP) and invited a number of distinguished scientists to join it. The aim of the AP was to

"to assimilate the scientific results from EUROTRAC and present them in a condensed form, together with recommendations where appropriate, so that they are suitable for use by those responsible for environmental planning and management in Europe"

Three areas were designated for study:

- photo-oxidants in Europe;

- acidification of soil and water and the atmospheric contribution to nutrient inputs;

- the contribution of EUROTRAC to the development of tools for the study of tropospheric pollution.

The first two stem directly from the scientific aims of EUROTRAC. The third was a recognition that much of what was done in EUROTRAC is of a long-term nature and serves to provide tools which will be required in the future to study the troposphere and so underpin policy development.

This volume presents the results of the Application Project drawing from the plethora of scientific results, the conclusions of EUROTRAC about the state of the troposphere over Europe. The various findings and recommendations on local, regional and global scales illustrate the complexities of the situation and indicate the consequent difficulties which face environmental policy makers seeking to find ways to abate pollution on a regional scale.

We recommend the book first for the scientific and policy-related information which it undoubtedly contains. In addition the book is an excellent illustration of how the scientific data and understanding from a comprehensive scientific programme can be evaluated in order to draw from it conclusions which are useful in policy development. That this has so successfully been achieved by the AP is a credit both to the its members and to EUROTRAC as a whole.

Clearly such work is not complete. Not only do many scientific uncertainties remain, but the recent developments in policy, with the emphasis on critical loads to ecosystems, require a still more quantitative approach to these problems on smaller scales. These coupled with the intrinsic complexities of the natural atmosphere will require much new understanding. It is hoped that this will at least in part be provided by the new project, EUROTRAC-2, recently accepted by EUREKA.

Many people, organisations and governments were involved, directly or indirectly, in the production of this book. Their contributions are fully acknowledged in Chapter 2 and within the general text. In offering our own thanks to all those, inside and outside EUROTRAC, who have contributed to its success we would like to thank in particular the members of the Application Project itself; the ultimate achievement of the project reflects their commitment and enthusiasm.

> Erik Fellenius, Swedish EPA,
> *Chairman of the International Executive Committee*
>
> Anthony R. Marsh, Imperial College, London,
> *Chairman of the Scientific Steering Committee*
>
> Wolfgang Seiler, IFU, Garmisch-Partenkirchen, Germany,
> *Director of the International Scientific Secretariat*

Chapter 1

Executive Summary

In this report, the scientific results from EUROTRAC have been assimilated by the Application Project, and the principle findings are presented in a condensed form, suitable for use by those responsible for environmental planning and management in Europe.

In most European countries, air pollution exceeds the acceptable national and international standards but, as many pollutants are transported great distances in the air, only international measures will be successful in control and abatement. The initial measures taken under the Convention on the Long Range Transport of Air Pollution (LRTAP) are bringing some benefits, but it is clear that further reductions in

| EUROTRAC and the |
| Application Project |

emissions will probably be very expensive to implement. Future cost effective abatement strategies will only be successful if they are underpinned by a thorough understanding of the detailed atmospheric processes in which the pollutants are produced, transformed, consumed, transported and deposited from the air. EUROTRAC, the policy-relevant findings of which are discussed in this report, was set up in 1985 to help provide the improved scientific understanding necessary for future policy development in the field of air pollution control and abatement. It was recognised that the scientific problems associated with trans-boundary air pollution could only be solved by an international high level state-of-the-art interdisciplinary project.

EUROTRAC is a co-ordinated environmental research programme, within the EUREKA initiative, studying the transport and chemical transformation of trace substances in the troposphere over Europe. The project consists of more than 250 research groups in 24 European countries and is organised into 14 subprojects. (section 2.4.1)

The EUROTRAC Application Project (AP) was set up

"to assimilate the scientific results from EUROTRAC and present them in a condensed form, together with recommendations where appropriate, so that they are suitable for use by those responsible for environmental planning and management in Europe".

Three themes were addressed by the AP and are presented in this report:

- Photo-oxidants in Europe; in the free troposphere and in rural and urban atmospheres;

- Acidification of soil and water, and the atmospheric contribution to nutrient inputs;

- The contribution of EUROTRAC to the development of tools for the study of tropospheric pollution; in particular, tropospheric modelling, the development of new or improved instrumentation and the provision of laboratory data.

The first two themes bear directly on issues of environmental concern in Europe. The third, "tools", is a recognition that, while some of the work done during the project has found use immediately, much will find its application in the longer term either directly, by serving applications to policy, or indirectly, by incorporation into the general understanding. (section 2.4.2)

1. The research accomplished in EUROTRAC provides substantial scientific support for the negotiations in the second generation of abatement strategy protocols under the UN-ECE Convention on the Long Range Transport of Air Pollution (LRTAP), in particular the second revised NO_x protocol and the revisions of the recently signed sulfur protocol. It also finds application for work within the European Environmental Agency, the EU Framework and ozone directives, and in the development of national strategies. EUROTRAC models and measurements are also used by WMO/UNEP and the IPCC for assessment of the current ozone budget and its sensitivity to changes in precursor concentrations.

> **Policy applications of EUROTRAC scientific results**

The application of effects-based control strategies by European governments, in which the maximum environmental benefit is being sought for the investment in control technology, places great demands on our knowledge and understanding of the links between sources, deposition and effects. Within EUROTRAC, major developments in the science for developing these links have been made. (Chapter 2)

Photo-oxidants in Europe: in the free troposphere, in rural and in urban atmospheres

> **The concentration of photo-oxidants in Europe is strongly influenced by photochemical production from man-made precursors that are emitted within the region**

2. From experimental studies it is concluded that the natural background of ozone over Europe at the turn of the century, within the atmospheric boundary layer, was about 10 to 15 ppb at ground level and 20 to 30 ppb one to two kilometres above the ground. Today, the concentration of ozone near the sea surface is 30 to 35 ppb before air masses move into Europe from the west.

On a seasonal basis, photochemical processes over western and central Europe add about 30 to 40 % to this background in summer, and subtract about 10 % in winter.

Within Europe very high concentrations of more than 100 ppb are observed during photochemical episodes under unfavourable meteorological conditions, *i.e.* high solar radiation combined with stagnant air or circulating wind systems.

In the free troposphere, that is from the top of the atmospheric boundary layer (1 to 2 km above the ground) to the tropopause which constitutes the boundary with the stratosphere (10 to 12 km above the ground), the background concentration before the air masses pass over Europe is higher than in the atmospheric boundary layer, being about 40 to 50 ppb in winter and autumn and 50 to 70 ppb in spring and summer.

The concentration of free tropospheric ozone over Europe is influenced not only by European emissions but also by North American and Asian emissions. (sections 3.2.1, 3.2.5, 3.3.2)

3. The concentration of ozone in the troposphere north of $20°N$ has increased since the beginning of modern measurements. This increase was larger at northern mid-latitudes than in the tropics, and larger over Europe and Japan than over North America. Comparison with historical data suggests that ozone in the troposphere over Europe has doubled since the turn of the century and that most of the increase has occurred since the 1950s. Measurements of nitrate in ice cores from Alpine glaciers provide strong circumstantial evidence for man-made emissions being responsible for the observed ozone trend. (section 3.2.1)

> **Tropospheric ozone in the northern hemisphere has increased since the 1950s**

4. Long-term observations show that the increase of ozone in the free troposphere was smaller in the eighties than in the seventies. The average ozone concentrations in the boundary layer near the ground have even decreased at some locations, for example at Garmisch-Partenkirchen in Germany and at Delft, a polluted site in the Netherlands. The concentration of peroxyacetylnitrate (PAN), a photo-oxidant like ozone, increased by a factor of three at Delft in the 1970s and stabilised in the 1980s. (section 3.2.1)

> **Tropospheric ozone increase slowed down in the 1980s**

5. The understanding of the temporal and spatial resolution of the emissions of NO_x and VOC has improved considerably. Results show for example that emissions are approximately 30 % lower at weekends than during the week. Such variations provide a regular "natural experiment" for examining the effects of the short-term reductions of emissions. The effects of the reductions on the photo-oxidant concentrations in this case appear to be rather small and indeed may increase ozone levels in areas of high pollution.

> **Weekday/weekend differences in the emissions of ozone precursors**

6. Recent model simulations have suggested that the effective abatement of elevated ozone concentrations in Europe require the reduction of the emissions of both NO_x and VOC, with more emphasis being put on VOCs, especially in north-west and central Europe. However some field experiments in EUROTRAC have identified possible shortcomings in the models that are presently used for quantifying ozone/VOC and ozone/NO_x relationships. These experiments emphasise the greater importance of NO_x emissions in controlling the photochemical ozone balance. The reasons are (sections 3.2.3, 3.2.5, 3.2.6, 3.3.2):

- There is an indication, based on ambient measurements, that biogenic VOC emissions, which cannot be subjected to abatement, form a base level of VOCs which is higher than that assumed previously.

  | **Should NO_x or VOC be controlled, or both?** |

- The photochemistry in urban plumes seems to proceed faster than is assumed in models. The results suggest that the oxidation of VOCs leads to more peroxy radicals and, hence, more ozone over a shorter time than predicted by the photochemical schemes currently used in atmospheric models, and to a faster removal of NO_x, the catalyst in ozone formation.

Based on today's knowledge, the following picture emerges:

- An effective way to reduce ozone concentrations on urban and suburban scales appears to be to reduce VOC emissions. However, NO_x reductions are required to reduce the concentrations of other oxidants such as NO_2 and PAN.

- Both NO_x and VOC reductions are required in order to reduce ozone levels on a European scale. NO_x reductions are essential to reduce ozone concentrations on a global scale.

- The emission reductions need to be substantial (40 to 60 %) to obtain noticeable reductions in ozone concentrations.

7. The pre-industrial concentration of ozone of about 10 to 15 ppb at ground level resulted from the approximate balance between the transfer of ozone from the stratosphere to the troposphere, the destruction in the troposphere by photochemical reactions and by deposition to the ground. The 15 to 20 ppb

  | **Are photo-oxidants a home-made or trans- boundary problem?** |

difference between today's ozone levels, near the sea surface before air masses pass over Europe, and the pre-industrial ozone concentration is probably due to photochemical formation from precursors such as VOC and NO_x, emitted in other parts of the northern hemisphere, in particular North America. Reduction in the background tropospheric level will require agreement on a hemispheric scale. In addition, precursors emitted from biomass burning have a large impact on ozone concentrations in the tropics and in the southern hemisphere. (section 3.2.6)

The very high ozone, NO_2 and PAN concentrations that are observed in some urban and suburban areas (photochemical smog) are due to photochemical production in the atmospheric boundary layer from precursors that are mostly emitted within the area. (sections 3.2.4, 3.2.5)

On a continental scale, the enhanced photo-oxidant concentrations observed are a consequence of both *in situ* chemistry and transport from regions with higher emissions. Studies in EUROTRAC have greatly added to our understanding of the relevant processes for quantifying ozone/precursor relationships and, hence, provide a better basis for determining how reductions in precursor emissions in one region would reduce the photo-oxidant levels in regions downwind, especially in moderately populated and rural areas (scales >50 km). However, source-receptor relationships are still difficult to assign because the chemistry is non-linear and there are large differences in the residence times of photo-oxidants and precursors in the atmospheric boundary layer close to the ground, compared with the residence times in the free troposphere. (sections 3.2.5, 3.2.6, 3.3.1, 3.3.2, 3.3.3)

While the highest photo-oxidant levels can be counteracted by local pollution control measures, abatement of enhanced ozone formation on a European scale requires a co-ordinated abatement strategy. (sections 3.2.4, 3.2.5, 3.3.1, 3.3.2)

8. Strategies to abate photochemical air pollution at any relevant scale may be assessed with models developed within EUROTRAC.

On the local scale, a zooming model (EZM) has already been successfully utilised to optimise the air pollution abatement strategy for Athens, to support the decisions taken with regard to traffic regulations in Barcelona during the 1992 Olympics and to interpret the observations during measuring campaigns in the Upper Rhine Valley.

Models have also been developed to describe the distribution of ozone in the global troposphere. The results show that the concentrations of ozone throughout large parts of the northern hemisphere have been substantially increased by anthropogenic emissions of nitrogen oxides, VOCs and carbon monoxide. Since ozone is a greenhouse gas, these elevated levels could be making an appreciable contribution to global warming. (section 3.3.3).

The EURAD model as well as global models have been used to calculate the influence of aircraft emissions on upper tropospheric ozone levels. (section 3.3.2)

Practical applications of photo-oxidant models on all scales

Acidification of soil and water and the atmospheric contribution to nutrient inputs

9. There is clear evidence that deposition of anthropogenic sulfur and nitrogen compounds over central and northern Europe has caused severe changes in the composition and functioning of many ecosystems. The deposition of these compounds has also been an important factor in the deterioration of materials and of our cultural heritage of ancient buildings. These effects have to a large extent occurred during the second half of the twentieth century and are mainly caused by emissions of sulfur dioxide, nitrogen oxides and ammonia *within* Europe. (sections 2.2.3 and 4.1)

> **Sulfur and nitrogen deposition is still causing severe damage to ecosystems in Europe**

   Annual budget calculations on a national scale and for Europe as a whole show that 50 to 70 % of the emissions of sulfur and oxidised nitrogen and approximately 80 % of the emissions of reduced nitrogen are deposited within Europe. These inputs exceed the critical loads for soils and for freshwater acidification over large areas of Europe. A protocol to the LRTAP for reducing the sulfur deposition over the next decade has recently been agreed. For nitrogen, protocols are being developed.

   SO_2 emissions vary markedly with the time of day, the day of the week and the season, the variations being explicable in terms of working hours and the influence of temperature on energy demand. (section 4.2.1)

10. The inputs of dimethyl sulfide (DMS) to the atmosphere are important on a global scale for the atmospheric sulfur budget. The emissions from oceans contribute to the background 'natural' sulfur, and sulfur compounds are deposited together with anthropogenically derived sulfur. However, except in small coastal areas of northern and western Europe, the contributions from natural sources to annual sulfur inputs is negligible. (section 4.3.1)

> **The natural sources of sulfur are unimportant relative to the anthropogenic sources in Europe**

11. The protocols agreed in 1994 should substantially reduce sulfur emissions and deposition throughout Europe by 2005, and a decrease in sulfur emission and deposition in Europe will reduce the scale of environmental damage due to acidification. Sulfur is however not the only contributor. Deposited nitrogen (both oxidised and reduced) contributes to the acidification problem and in many areas the deposition of nitrogen alone exceeds critical loads to ecosystems. The downward trends in sulfur emission over the last decade, and those expected in the next, are rapidly increasing the relative importance of nitrogen in acidification and eutrophication. (section 4.3.3)

> **The contribution of deposited nitrogen to acidification and eutrophication is increasing**

12. Cloud chemical processes have been shown to be at least as important as gas-phase processes for the oxidation of sulfur dioxide. New pathways for the aqueous-phase oxidation have been discovered and quantified. Incorporation of transfer and

> **Most of the sulfur dioxide oxidation in Europe occurs in clouds**

reaction mechanisms in simulation models is in progress and it should soon be possible to include more realistic cloud modules in source-receptor models. (section 4.2.2)

13. Field experiments performed in recent years have led to a better understanding of the formation of cloud droplets and of how particulate matter is incorporated in the droplets. In these experiments two groups of aerosols were observed with different hygroscopic properties. These findings show that the different

> **Cloud chemical processes cause non-linearity**

chemical species in aerosols are scavenged by clouds and therefore by wet deposition processes at different rates, which will lead to different residence times in the atmosphere. (section 4.2.3)

The reduction of sulfur emissions in western Europe after 1980 is not reflected in atmospheric concentrations in a simple way. Measurements show that sulfur dioxide concentrations are reduced faster than expected, while concentrations of sulfur in precipitation show a slower decrease than expected. Aqueous-phase chemistry in clouds may be a key factor contributing to the observed anomalies. (section 4.2.3)

14. The application of approaches to develop emission controls, based on critical loads, provides a means of maximising the environmental benefits for the investments in controls. These approaches require the ecosystem sensitivity and the actual deposition input to be quantified on the same scale. This scale of variability in ecosystem sensitivity to acidification

> **Effects-based emission controls demand great spatial detail for ecosystem sensitivity and deposition**

varies with land use and vegetation, with the landscape scale being mostly between 1 and 10 km. (section 4.3.4)

15. Models and data bases have been developed to provide the fine scale (1 to 10 km) resolution in the land-use-specific deposition maps that are required to calculate exceedances of critical loads. These methods have been developed from field and laboratory based studies of deposition processes and validated by long-term flux measurements. The new equipment developed to make the measurements, the collaborative field campaigns and the long-term flux studies have provided the science which

> **Fine-scale deposition maps of the key acidifying species have been provided by EUROTRAC work**

underpins the deposition maps. The work has been developed for both national and international approaches.(section 4.3.3)

16. The dry deposition of sulfur dioxide to vegetation has been shown to be influenced by the presence of surface water (rain, dew *etc.*). These effects lead to larger rates of SO_2 deposition in the presence of ammonia. Atmospheric ammonia may therefore reduce the atmospheric lifetime of sulfur dioxide. Similar effects of SO_2 on rates of ammonia exchange over

> **The elimination ammonia emission Europe would cha the trans-bounda fluxes of sulfur**

vegetation are expected. The quantitative detail has yet to be provided, but it is clear that emission controls of either of these gases will influence the lifetime and the deposition 'footprint' of both gases. (section 4.2.3)

17. Up to now the control of emissions of nitrogen compounds has been largely on emissions of NO_x from vehicles and large combustion plants. However some of the major environmental effects (acidification and eutrophication) are caused by *both* oxidised and reduced nitrogen. Moreover, deposition and the

> **Reduced as well oxidised nitrog emissions need t controlled**

effects of deposited nitrogen over large areas of Europe are dominated by inputs of reduced nitrogen as NH_3 in dry deposition and NH_4^+ in wet deposition. To reduce the effects or eliminate exceedances of critical loads for nitrogen, it is necessary to control emissions of both NO_x and NH_3. (section 4.2.1 and 4.3.1)

18. Models have been developed to describe the global transport of sulfur and nitrogen compounds. These models show that sulfate aerosols (originating from anthropogenic sulfur emissions) reflect sunlight back into space and thereby cool the surface. In heavily industrialised regions,

> **Anthropogenic sulfa aerosols in the fre troposphere counter global warming**

where the concentration of sulfate aerosols is high, this cooling effect may mask a major fraction of the warming due to greenhouse gases. (section 4.3.5)

The contribution of EUROTRAC to the development of tools for the study of tropospheric pollution

19. Methods and tools, in the form of simulation models, instrumentation and the results of laboratory studies, have been developed which are being used and will be used in the future in the evaluation of abatement strategies on continental and urban scales, for air pollution episodes and long-term averaged situations. (section 5.1)

> **Tools now availabl for the evaluation o abatement strategie**

20. Major improvements and new developments in instrumentation have been made. These include long-path absorption techniques, tunable-diode laser spectrometers, tropospheric lidar techniques, automated VOC measuring systems and a sensitive NO_x analyser. Numerous instrument intercomparisons have led to a better assessment of the accuracy of the instruments. Several of these techniques are now in general use and have been

> **New and improved instrumentation ha been developed with this EUREKA proje**

commercialised; as expected in a EUREKA project. The EUROTRAC experience indicates however that if the *separate* development of innovative instrumentation is required, precisely defined objectives and centralised funding are needed. (section 5.2.5)

21. Laboratory studies of free radical processes, of the oxidation reactions of aromatic and biogenic compounds and of individual reactions of many chemical intermediates have considerably improved the knowledge of tropospheric ozone chemistry. Laboratory studies of chemistry in cloud droplets and aerosol particles have added to the understanding of the elementary reaction steps of the oxidation of sulfate in clouds, of the absorption process of gases by cloud droplets and of aerosol surface reactions. These new findings have been incorporated into simulation models, so improving their reliability. (sections 5.3.2, 5.3.3 and 5.3.4)

> **Laboratory results are improving our knowledge of atmospheric chemistry**

22. Simulation models on urban, continental and global scales have been developed, improved and validated. These models, which integrate the results of process-oriented studies, describe in a quantitative manner the relationship between emissions and atmospheric concentrations or deposition. As such models are based on state-of-the-art knowledge and experience, and they should be viewed as adequate and reliable tools which have been and will be asked to answer policy-oriented questions. However simulation models require continuous improvement to include new findings and understanding, and frequent validation to demonstrate their reliability. (section 5.4.7)

> **Simulation models required to address policy-oriented questions**

23. Emission data bases, including the procedures developed for temporal and spatial disaggregation of the data, are an essential tool since both accurate emission data and optimal meteorological input is required as input for simulation models in order to provide quantitative source/receptor relationships. Only with good and accurate emission data can reliable results be expected from simulation models. Data bases of field observations and field campaigns collected within EUROTRAC are also valuable tools for future studies. (sections 3.2.2, 5.5 and 5.5.2)

> **Good emission data essential**

24. The network of scientists, created by EUROTRAC, has and will lead to more synergistic scientific activities and a coherent environmental policy in Europe. The network provides an efficient way of transferring knowledge and provides a major source of expert advice to the participating countries, to the EU and the UN-ECE. (section 5.5.5)

> **Networks of scientists are sources of expert advice**

Uncertainties in our present knowledge

25. While much has been achieved within the present project a number of uncertainties remain. These will have to be resolved in the future if the source/receptor relationships, quantitative enough to support the present trends in policy development, are to be produced. The following headings mention the major uncertainties and lists some of the work needed to reduce them.

> **Uncertainties requiring resolut**

For photo-oxidants, there are uncertainties in estimates of precursor emissions and in the detailed chemical and meteorological mechanisms by which photo-oxidants are processed. Studies are necessary to:

- develop better emission estimates for nitrogen oxides and speciated volatile organic compounds including biogenic species;

- validate chemical transport models on regional and sub-regional scales;

- resolve the current disagreements between field experiments and model calculations on the role of NO_x versus VOC limitation of photo-oxidant formation.

For acidification and deposition of nutrients, the principle uncertainties are in emission estimates, the effects of clouds and aerosols on the production of acidity and the effects of complex terrain on deposition. Studies are necessary to:

- develop better validated emission estimates for nitrogen oxides and ammonia;

- develop a more complete understanding, and hence parameterisation of aerosol processes governing their life cycle from their formation and transport to their transformation and deposition in the atmosphere, and incorporate this into source-receptor models;

- develop validated models for local-scale deposition of reactive species;

- develop methods and models to quantify deposition in complex terrain (hills and forests).

For the development of tools, improvements are needed in the models themselves, in their validation, in instrumentation for monitoring and research and in the mechanisms of the fundamental reactions involved in the formation of photo-oxidants and acidity. Studies are necessary to:

- validate models with more dedicated field measurements and specifically designed field campaigns;

- validate emission data using an integrated approach that includes ambient concentration measurements and the use of models;

- increase and improve scientific monitoring for process studies, trend analysis and the evaluation of policy measures;

- develop improved instrumentation for monitoring and field measurements within any future scientific programmes;

- elucidate the mechanisms for the oxidation of aromatic and biogenic VOCs.

- complete the chemical mechanisms used in models and test them in chemical simulation experiments; particularly requiring attention are the mechanisms describing the effect of aromatic hydrocarbons on photo-oxidant formation, the oxidation of biogenic VOCs, and nitrogen chemistry, including heterogeneous processes;

- incorporate validated and more realistic mechanisms for aqueous oxidation in cloud droplets into simulation models.

Chapter 2

Environmental Problems, Policies and EUROTRAC

2.1 The atmosphere as a recipient and transport medium for pollutants

The atmosphere is a main recipient for gaseous waste from the industrialised society. The combustion of fuels for heating, electricity production, industrial processes and transportation is the most important source of atmospheric pollutants. Both industrial manufacturing, and agriculture also cause significant atmospheric emissions.

In the atmosphere, pollutants are diluted during transport from the sources to remote, cleaner areas. In the past the capability of the atmosphere to dilute emissions has made air pollution acceptable and it was only when sources became large and dense that serious effects were recognised. Most of the regional transport and deposition proceeds within the atmospheric boundary layer (ABL). The ABL atmospheric boundary layer is the lowest part of the troposphere and extends to between 500 and 2000 m from the ground. On average, atmospheric emissions are transported at a speed of 20 to 40 km h^{-1} so that an air parcel will remain over an area of the size of an average country less than 24 hours. Since the mean residence time of pollutants in the atmosphere often exceeds this time, air pollution is a transboundary problem.

The atmosphere also acts as a chemical reactor, in which the emitted pollutants undergo chemical conversions. Many compounds, insoluble in water, become soluble on oxidation and may then be easily removed from the atmosphere. Chemical reactions convert organic compounds to carbon dioxide and water, and also form compounds such as photochemical oxidants that are harmful to man and to the environment.

The atmosphere plays a crucial role in the distribution and impact of air pollutants. Understanding and quantifying the relationships between sources and receptors, and the effects occurring at the receptors have become one of the most important tasks in the development of abatement strategies, and for national and international decisions on the control of the emissions.

The scales involved in the abatement of air pollution span from the controlling of the distribution of air pollution on a local scale (*e.g.* an urban area) to air pollutants distributed all over the globe. Here a classification is made in terms of spatial scale of distribution and impact:

- Urban or local issues have a horizontal scale of 10 to 100 km, a time scale of hours up to one or two days, and are confined to the atmospheric boundary layer. Abatement of local and urban air pollution can be achieved within the local and urban areas.

- Regional and continental issues have a horizontal scale between the local or urban scale and up to 1000 to 5000 km, a time span of a few days, and are still confined to the atmospheric boundary layer. Abatement of regional or continental air pollution can be achieved within the regional or continental scale.

- The global scale covers the troposphere and the stratosphere, the whole globe and is characterised on a time scale extending from one week and upwards to many years. Abatement of global pollution requires global scale actions. (Table 2.1)

The control of effects due to atmospheric emissions is based on the assumption that there is a link between emissions and effects and that reductions in emissions will decrease or eliminate the effects. Looking at how regional air pollution problems have been handled, three different stages in establishing a basis for control measures can be discerned;

- a discovery stage,

- a stage of common understanding and

- a stage where quantitative source-effect relationships may act as the basis for control measures.

During the first stage, emission control is sometimes questioned and emission control measures are sporadic. During the second stage, the stage of common understanding, joint measures are taken, but regulations and agreements are generally limited to simple percentage reductions. The principles for control are often based on best available technology (BAT) and/or precautionary principles. In the third stage, the stage of quantitative source-effect relationships, control strategies become more sophisticated and it should be possible to predict the efficiency in different control measures.

The various atmospheric environmental problems have reached different stages. Acidification is the most developed, and quantitative source-receptor relationships have been used in the recently signed second protocol on sulfur emissions under the *UN-ECE Convention on Long-range Transboundary Air Pollution* (UN-ECE, 1994).

Table 2.1:   Characteristic transport distances in the vertical and horizontal scales and the atmospheric lifetimes of SO_2, NO_2, NH_3, and VOC and their transformation products (sulfate and nitrate aerosols, ozone *etc.*).

	Vertical scale (km)	Horizontal scale (km)	Time scale
Acidification	2	2500	10 days
Eutrophication	2	1000	5 days
Photo-oxidants			
Local	2	100	1 day
Episodic	2	2500	10 days
Growing season	2	2500	100 days
Global	10	20000	100 days
Climate change			
Carbon dioxide	global	global	100 years
Methane	global	global	10 years
Nitrous oxide	global	global	100 years
Ozone	10	5000	100 days
Sulfate aerosols	5	5000	10 days
Stratospheric ozone layer			
Polar ozone	10–15	polar	< 100 days
Global	10–30	global	100 years

2.2   The environmental problems

2.2.1   Environmental problems covered within EUROTRAC

The EUROTRAC project focused on regional air pollution problems with their main distribution scale within Europe and adjacent areas. It has to some extent also covered urban air pollution problems, in the Athens area, for example, and the links between regional and global scale problems, such as the formation and transport of ozone in the free troposphere. The main focus of EUROTRAC has however been on the formation of oxidants in the atmospheric boundary layer and the transport and deposition of pollutants causing acidification and eutrophication. Table 2.2 lists the main pollutant groups of interest within EUROTRAC and the related environmental problems. Also listed are the main physical and chemical processes which, in addition the emissions themselves, are of importance for the impact at the receptors. The table also gives an indication of the level of attention paid in EUROTRAC to research into the processes that determine transport, transformation and removal.

Table 2.2:   The main pollutants of interest within EUROTRAC and the related environmental problems. The table also indicates the main physical and/or chemical processes which, in addition to the emissions themselves, are of importance for the impact at the receptors. In the last column is given an indication of the level of attention paid in EUROTRAC to research into the processes that determines the receptor impacts.

Environmental problems	Processes of importance	EUROTRAC relevance
Sulfur		
Health – SO_2	distribution	low
Health – SO_4	chemistry, distribution	low
Materials – SO_2	distribution	low
Visibility	chemistry, distribution	medium
Acid deposition	chemistry, distribution, deposition	high
Climate		
clouds, fog, precipitation	chemistry	high
radiation	distribution, chemistry	medium
Nitrogen (NO_x and NH_3)		
Health – NO_2	oxidation, distribution	low
Oxidant formation NO_x	chemistry, meteorology	high
Acidification NO_x and NH_3	distribution, chemistry, deposition	high
Nutrient effects NO_x and NH_3	distribution, chemistry, deposition	high
VOC		
Health	distribution	low
Oxidant formation	natural emissions, chemistry, meteorology	high
Oxidants and oxidant formation		
Health – ozone	chemistry, distribution	high
Effects on vegetation	chemistry, distribution deposition	high
Effects on materials	chemistry, distribution	high
Climate effects	chemistry, distribution	high
Change of "oxid. capacity"	chemistry	high

Regional air pollution problems are closely connected in several ways and to reduce the impacts of one regional pollution effect, often more than one pollutant needs to be controlled. Moreover, measures taken to control any of the regionally distributed pollutants may affect other environmental problems, not only on a regional scale, but also on local and global scales. The interconnections between chemical compounds, their sources, effects and receptors for pollution problems occurring on a regional scale are illustrated in Fig. 2.1 (Grennfelt *et al.*, 1994).

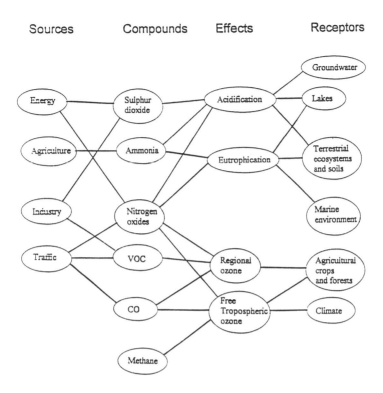

Fig. 2.1: The regional air pollution problems in Europe. Relationships between dominant sources, emitted compounds and their effects on different receptors. Other links may be established, for example from ozone to health effects. (Grennfelt *et al.*, 1994)

2.2.2 Photochemical oxidants and their effects

Photochemical oxidants are secondary pollutants formed in the atmosphere under the influence of sunlight. The precursors of oxidant formation are volatile organic compounds, VOCs, carbon monoxide and nitrogen oxides. VOCs and CO act as "fuels" as they are oxidised by the process and the nitrogen oxides as "catalysts", since they are not consumed directly in the oxidation. The chemistry is further discussed in Chapters 3 and 4. The most important product is ozone and the main environmental effects due to photochemical oxidants are caused by ozone.

Photochemical oxidant formation occurs on various scales, from local, as in urban areas such as Los Angeles, Athens and Milan, and regional, as in central and north-western Europe, to (semi) global, as the increase in background concentrations over mid-latitudes in the northern hemisphere. Locally formed oxidants generally show large temporal and spatial variations with high peak concentrations caused by emissions mainly occurring the same day. Regional oxidant formation most often occurs in connection with stable high pressure zones and high concentrations may remain for several days.

Locally formed photochemical oxidants, in particular ozone, have for decades been known to cause acute health problems and have severe effects on agricultural crops. Regional episodic formation of ozone and other photochemical oxidants in Europe was discovered in the early 1970s (Atkins *et al.*, 1972; Cox *et al.*, 1975). It was also recognised that this is more often the case in northern and western Europe than in the south of the continent, where photo-smog is often of a local character (Bouscaren, 1991).

The ozone episodes were shown to damage agricultural crops, such as spinach and potatoes, for which typical symptoms were visible injuries to the leaves. Several important crops, particularly wheat, were also found to be affected by ozone, but the effects primarily appear as decreased yields often without visible injuries to the plants. The present oxidant situation is expected to cause crop losses of the order of 10 % in polluted areas in central Europe.

There has also been an indication that ozone contributes to the forest decline in central Europe (Ashmore *et al.*, 1990). Controlled experiments have shown that present ozone concentrations may cause decreased photosynthetic production and probably also decreased forest production. The ozone concentrations observed and their possible threats to man and environment have played a central role in air pollution control measures, for example in the UN-ECE protocol on VOC.

As indicated in Chapter 3, in summer there is a mean enhancement of 10–15 ppb of ozone in some areas in central Europe, while in winter there is a decrease in ozone where local factors like emissions of nitrogen oxides and dry deposition play a role. Ozone episodes with hourly peaks of 100 ppb and more occur frequently over central Europe every summer. In the northern UK and Scandinavia, summer episodes seldom give concentrations above 100 ppb. In the Mediterranean, local and mesoscale oxidant formation may be more important than regional transport (Fig. 2.2a).

There is strong evidence for a doubling in the annual average boundary layer ozone concentration over Europe since the turn of the century (see section 3.2.1), and in order to protect human health, crops, ecosystems and materials, air quality standards and guidelines have been developed. The WMO has set guidelines for the protection of human health of between 75 and 100 ppb as maximum hourly exposure values. The UN-ECE Task Force on Mapping has also suggested a critical exposure level for ozone.

Tropospheric ozone is an important greenhouse gas and the increased background concentrations are expected to make an appreciable contribution to the radiation balance. The increased background concentrations also influence the chemistry of the troposphere and thus the oxidation capacity of the atmosphere; they also add to the ozone produced on local or regional scales. There are even indications that the increased background concentrations over the northern hemisphere may themselves have negative effects on vegetation.

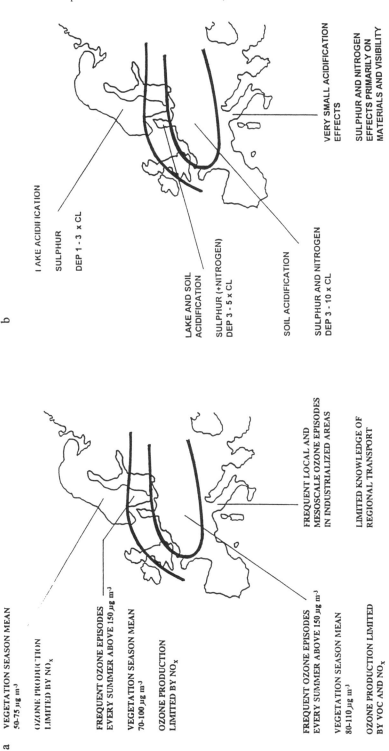

Fig. 2.2:   Air pollution regions in Europe. a) Photochemical oxidants; b)Acidification and eutrophication. CL denotes critical loads. (Grennfelt *et al.*, 1994)

2.2.3 Acidification and eutrophication

The acidification of soils and surface waters is a major environmental problem in Europe. Acidification of precipitation and surface waters was discovered in 1967 and recognised as a large-scale air pollution problem requiring international actions for its solution (Odén, 1968). Biological effects in terms of the disappearance of fish populations were discovered in Scandinavia at about the same time (e.g. Hultberg and Stensson, 1970; Jensen and Snekvik, 1972).

Acidification is mainly caused by the deposition of sulfur and nitrogen compounds, originating from emissions of sulfur dioxide, nitrogen oxides and ammonia. The main sources of sulfur and nitrogen oxides are combustion of fossil fuels. Ammonia is primarily emitted in agricultural activities; it is a neutralising agent in the atmosphere but on reaching the ground it is easily converted to nitrate and may in this way cause acidification. Soil acidification has been observed over large areas in northern and central Europe. In Scandinavia sulfur deposition is the main cause for the acidification, while in central Europe nitrogen deposition also makes an important contribution (Table 2.3, Fig. 2.2b). Field studies have shown that glaciated soils in northern Europe have lost a large part of their base saturation and many soils show pH values between 0.5 and 1 pH units less than they were during the first half of the 20th century (Hallbäcken and Tamm, 1986). Soil acidification is assumed to be one of the most important factors for the forest decline observed in many areas in Europe. The acidification of soils gives high concentrations of inorganic aluminium in the soil solution which is thought to be toxic to tree roots (Ulrich and Matzner, 1983).

Table 2.3:   The relative importance of sulfur and nitrogen for the soil acidification in different European regions (Grennfelt et al., 1994).

Region	Input keq/ha yr	Input keq/ha yr	Output keq/ha yr	Output keq/ha yr	N fraction in output
	(SO_4^{2-})	(NO_3^-/NH_4^+)	(SO_4^{2-})	(NO_3^-)	(%)
Northern & central Scandinavia	0.3	0.2	0.3	< 0.05	< 15
Southern Scandinavia	1–1.5	0.7–1.5	1–1.5	0.1	7–10
Central Europe	2–4	1–3	2–4	0.3–1	10–30
The Netherlands	2–4	3–6	2–4	2–6	30–80

Lake acidification is an important environmental problem in the northern parts of UK, Norway, Sweden, Finland and the adjacent areas of Russia. In many lakes fish reproduction has disappeared and thousands of lakes in Scandinavia are treated with lime in order to preserve viable fish populations. Ground water acidification has led to the corrosion of water supply systems in Scandinavia.

Nitrogen compounds deposited from the atmosphere act, in addition to their role in acidification, as nutrients. The growth of many terrestrial and marine ecosystems is naturally limited by nitrogen; in such systems atmospheric deposition will act as a fertiliser. In the beginning nitrogen deposition stimulates growth and overall production. The observed increase in tree growth in parts of Europe is assumed to be at least partly due to the deposition of nitrogen compounds. Increased deposition then successively changes the nitrogen-limited ecosystems towards more eutrophic systems, where nitrogen-favoured species are more abundant. Finally, the systems may be so rich in nitrogen that their stability is threatened (Gundersen, 1992). Nitrogen deposition has changed vegetation composition in many areas in north-western Europe, especially in the Netherlands (Bobbink *et al.*, 1992). The effects are most pronounced in areas with intense animal production, where ammonia emissions from manure handling are the main cause of the deposition. A major effect is the changes in ecosystem composition, particularly in the Netherlands, where large areas of caluna heathlands have been converted to grasslands. High nitrogen deposition may also cause severe damage to forests close to areas with intense animal production (Innes, 1993). It has also been suggested that nitrogen deposition may be one important reason for the more widespread symptoms of ecological damage occurring over large areas of Europe.

In marine systems, nitrogen input from atmospheric deposition, rivers and direct waste-water discharges increases the frequency and magnitude of algae blooms. Such blooms may lead to oxygen depletion in the bottom water, when the algae after death sediment to the bottom. Severe algae episodes have been observed in coastal seas around the European continent (*e.g.* Berg and Radach, 1985; Andersson and Rydberg 1988). For the Baltic and the North Sea, budget estimates indicate that direct atmospheric deposition is responsible for approximately 30 % of the total input of nitrogen. In addition, some of the riverine input of nitrogen may be caused by atmospheric deposition being leached from various ecosystems.

In addition to acidification and nutrient effects, sulfur and nitrogen emissions may cause effects to human health and materials, primarily by the action of sulfur dioxide and nitrogen dioxide.

Photochemical oxidants normally do not interact in a simple way with acidification and eutrophication, since oxidants primarily affect the canopy while acidification affects the soil. There are, however, a number of investigations showing synergistic effects between ozone and SO_2 and NO_2 (*e.g.* Guderian and Tingey, 1987). The combined stress from acid deposition and photochemical oxidants may increase the potential for forest damage. Acid fog deposition has been shown to interact with ozone, causing increased leaching from coniferous canopies. There have only been a limited number of studies of the interactions between different compounds so that results and hypotheses are preliminary.

2.2.4 Links to global effects

Photo-oxidants and acidic substances may contribute to climate change effects in various ways. The direct effect from tropospheric ozone has already been mentioned together with the role of ozone in changing the oxidation capacity of the atmosphere and thus the lifetime of gases influencing the atmospheric radiation budget. There are also other ways, by which pollutants considered primarily as regional may contribute to the global change problems.

The formation of sulfuric acid, particulate sulfates and nitrates in the atmosphere, may cause a significant contribution to visibility reduction in polluted areas. Sulfate aerosols may also influence climate; either directly through reflection and absorption of solar radiation, or indirectly by changing the absorption and reflection properties of clouds as well as the lifetime of clouds. Aerosols have a negative effect on the radiation balance of the earth and thus counteract global warming. Model calculations show the importance of this effect (Charlson *et al.*, 1991).

Acidification and eutrophication may contribute to climate change by influencing the exchange of gases between soil and atmosphere. The most discussed effects are: an increased emission of nitrous oxide due to the high nitrogen deposition to terrestrial and aquatic ecosystems; and a decreased methane emission from wetlands due to the acidic deposition. The overall importance of these effects still needs to be determined.

Climate change and increased UV radiation due to depletion of the stratospheric ozone layer may in some cases increase ozone formation. The interference may be in two directions. One is the influence of global change on the regional effects and the other the influence of regional air pollution on climate change.

Acidification effects can be enhanced by climate change. Increased soil temperatures are expected to increase mineralisation and thus increase the leaching of nitrogen and sulfur from soils. Experimental evidence of the magnitude and relative importance is, however, lacking. Increased UV radiation due to depletion of the ozone layer may interfere with the direct effects of ozone on plants, although this effect has not been demonstrated experimentally. The effect is, however, plausible since tropospheric ozone and UV radiation do affect plant cells in a similar way.

2.3   Policies

In this section the development of policies for regional and global air pollution problems will be discussed. It will focus on the international policies and actions. The description is based on the three stages mentioned earlier; discovery, common understanding and quantifying source-receptor relationships. (Fig. 2.3)

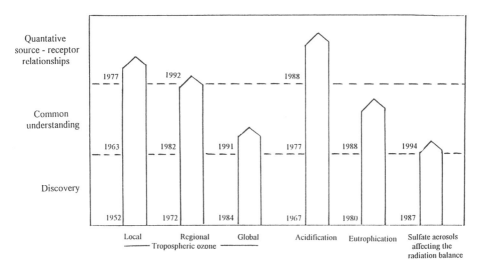

Fig. 2.3:   The development of the environmental problems studied within EUROTRAC. The years in the figure are only indicative and there is in most cases no single year, when environmental problems were discovered or turned from one level to another.

Policies require a number of tools developed by scientific research. These include

-   monitoring methods and networks in order to detect and understand the processes involved, verify models and determine trends in concentrations and deposition,

-   models in order to link effects (or receptor concentrations or processes) to emissions, and

-   models to predict the consequences of changes in emissions or the required emission reductions to achieve certain environmental goals.

The link between environmental science and policy is often not straight forward. Since environmental problems have been of increasing concern within society, decisions must often be taken on limited scientific evidence and knowledge.

Scientific evaluation and assessment are tools in the process of developing abatement strategies. The procedure of scientific assessment has been developed in a consistent way within the IPCC and the Vienna Convention for the protection of the ozone layer for the protection of the stratospheric ozone layer, which regularly publish assessment reports. For acidification and photochemical oxidants, the procedure is less developed, although assessment reports covering certain issues are published frequently. When knowledge is limited, the decisions to be taken have to be based on risk assessments, which often give the choice between taking action today on the basis of limited evidence, or waiting until the evidence is clearer by which time a more optimal decision may be taken but valuable lead time lost.

2.3.1 Policies used to combat regional air pollution problems

a. Early strategies on photochemical oxidants and acidification

The discovery stage for local photo-oxidant formation developed when it was found that urban emissions were a source of photo-oxidants in the Los Angeles smog in the 1950s (Haagen-Smit, 1952). Originally, photochemical air pollution was thought to affect primarily urban areas. The first policy actions to combat photochemical oxidants were also solely directed towards the urban areas, especially the Los Angeles area. In Europe, local oxidant production was observed about 1965 and measures were taken, especially in the Netherlands. In the 1970s, it was found that photochemical episodes could be associated with long-range transport of ozone and its precursors. Using simulation models, it was shown that the anthropogenic emissions of nitrogen oxides and volatile organic compounds could give ozone concentrations far above the background concentrations. The models also indicated that ozone or its precursors could be transported over long distances in the atmospheric boundary layer (Hov *et al.*, 1978; Eliassen *et al.*, 1982).

A level of common understanding of the distribution, and sources of local, long-range transported and global tropospheric ozone has evolved over the past decade; an assessment of the contribution by EUROTRAC is given in Chapter 3 of this report.

The policy application of the scientific understanding has come through national and international air quality standards, for example WHO guidelines and the EU directive for ozone, and emission regulations for the precursors, for example the *Protocol on Volatile Organic Compounds* (VOC) within the Convention on the Long-Range Transport of Air Pollution (LRTAP) (Table 2.4).

The LRTAP Convention deserves further consideration due its central role in the control of regional air pollution problems. Almost all the countries in Europe, as well as Canada and the United States, are members of the Convention. During the 1980s the LRTAP was an important bridge between the eastern and western countries in establishing a joint basis for environmental legislation, as well as for a common view on acidification, photochemical oxidants, heavy metals and other transboundary air pollution problems. Within the Convention, an institutional framework was set up bringing research and policy together. The main part of this framework is the *Co-operative programme for Monitoring and Evaluation of the Long-range Transport of Air Pollution in Europe* (EMEP). The programme, originally established in 1977, became a protocol to the Convention in 1984. The protocol regulates the financing of EMEP. The programme has three main elements; collection of emission data, measurements of the constituents in air and precipitation, and modelling of atmospheric distribution and deposition. EMEP is scientifically based and has been continuously improved by making use of new scientific results from various research programmes including EUROTRAC. Results from EMEP have played a significant role in the process leading to the protocols on emission reductions of air pollutants in Europe. The results of EMEP model calculations of transboundary air pollution played a particularly important role in the negotiations for the 1994 protocol on the further reduction of sulfur emissions. This protocol is based on the *critical loads approach*, in which the transfer matrix calculated by EMEP is an important element. An application of results from EUROTRAC within EMEP would thus be a clear illustration of EUROTRAC's contribution in solving environmental policy questions. Within EMEP there are three centres, a Chemical Co-ordination Centre in Lillestrøm, Norway; and two synthesising centres, in Oslo and in Moscow. The EMEP network consists of more than 100 monitoring stations in 32 European countries. In addition to EMEP, there are co-operative programmes on effects and on abatement technologies. The Convention has its secretariat in Geneva.

Table 2.4:   Signed protocols under the LRTAP Convention. The main objective of the VOC protocol within LRTAP is to control photochemical oxidant formation and effects, although it also mentions health effects as a factor.

Protocol	Dates	Content	Signatures May 1994
EMEP	Adopted in 1984, entry into force in 1987	Monitoring and assessment of transboundary air pollutants	Ratified by 35 parties
Sulfur	Adopted in 1985, entry into force in 1987	30% reduction in emissions between 1980 and 1993	Signed and ratified by 21 parties
Second sulfur protocol	Adopted in 1994	Effect-oriented cost-effective reductions based on critical loads	Signed by 28 parties in June 1994
Nitrogen oxides	Adopted in 1988, entry into force in 1991	1994 emissions should not exceed those in 1987*	Signed by 26 and ratified by 24 parties
Volatile organic compounds	Adopted in 1991	Emissions should be reduced by at least 30 % by 1999 with any of the years 1984–1990 as base year	Signed by 23 and ratified by 8 parties

*   Twelve countries have agreed to reduce their emissions by 30 % by 1997, with a base year between 1980 and 1986.

For acidification, the discovery stage can be roughly defined as the period from 1968 until about 1977, when the first exploratory project on transboundary air pollution, set up by OECD in 1972, was finalised (OECD, 1977). The OECD report reflects a common understanding between the European OECD countries that sulfur is a transboundary air pollution phenomenon. During that period some countries took measures to control sulfur emissions to alleviate the acidification problem. However, several countries increased their emissions while others introduced control measures primarily to reduce the effects of sulfur dioxide on health.

In the late 1970s, the acidification problem entered a second phase characterised by the existence of the common understanding that sulfur emissions caused a transboundary air pollution problem and that international efforts were necessary to solve this. The second phase lasted about 10 years. Two international agreements appeared during this period: the Convention on Long-Range Transboundary Air Pollution (LRTAP) signed in 1979 and the *Sulfur Protocol* to that convention signed in 1985 (Table 2.4). National control strategies were also developed in an increasing number of countries. Germany, for example, implemented substantial reductions to their sulfur and nitrogen oxide emissions by enforcing strict controls on large emitters. The CEC directive on large combustion plants was passed at the end of this phase as a consequence of the common understanding of the environmental problems that the emissions of sulfur and nitrogen oxides generate (CEC, 1988). Nitrogen oxides and ammonia were not part of the international strategies at this time and emissions in most countries increased. The first *Protocol on Nitrogen Oxides*, signed in 1988, did not include a real reduction in emissions; it only stated that 1994 emissions should not exceed those in 1987. This protocol has in fact been difficult to fulfil for countries in which emissions are expected to increase during the period. The national control measures introduced for large combustion plants in some countries (primarily Germany) and the early regulations on NO_x emissions from cars only served to slow the increase in emissions; they did not reduce them.

b. Critical loads and levels

In the late 1980s the policy work entered into the third phase. By this time a change in attitude towards environmental problems had appeared. In 1987, the *Brundtland Commission* published their report *"Our Common Future"* in which the view of sustainable development was a leading theme (UN, 1987). It was suggested that environmental pollution should not be allowed to reach levels at which ecosystems would be threatened. The critical load concept was introduced at the same time as means for developing control strategies (Nilsson and Grennfelt, 1988). It fitted well into the view of sustainability since it defined the pollution loads and levels below which no harmful effects would appear, even in the long term. The critical load was defined as

"A quantitative estimate of an exposure to one or more pollutants below which significant harmful effects on specified sensitive elements of environment do not occur according to present knowledge".

In 1988, the Executive Body within the LRTAP Convention agreed that further emission reductions of transboundary air pollution should be effect-oriented (*i.e.* based on critical loads) and thus involve a quantification of the source-effect relationships. In the second LRTAP protocol on further reductions of sulfur emissions in Europe, which was signed in June 1994, the quantitative relationships between sources and effects have, for the first time, been used to develop the most cost-effective control strategy (UN-ECE, 1994). During the last five years, the critical-load approach has been accepted by most countries in Europe as a means of quantifying the areas of environmental damage and in guiding policy to maximise the environmental benefits of control strategies for the investments made.

The inclusion of nitrogen compounds as contributors to acidification has been to a large extent parallel, but mostly with a time lag of a couple of years. So the first protocol on the control of nitrogen oxides was signed in 1988 and the critical load concept not fully developed until 1992 (Grennfelt and Thörnelöf, 1992). Negotiations on a second protocol under the LRTAP Convention, which started in 1994, are also based on the critical load/level concept. In the preparation for the new protocol, there is also a general agreement between the parties under the convention to take a further step into effect-oriented approaches by including several effects and several compounds within one protocol.

In an analogous way to the critical loads, the concept of critical levels was developed to handle the direct effects of pollutants (ozone, sulfur dioxide *etc.*) on vegetation and materials, and a preliminary set of levels were set in 1988 (UN-ECE, 1988). The first set of critical levels for ozone were revised in 1992 and in 1993, introducing a basis, where the exposure above a certain threshold level during the vegetation season was introduced as a way of expressing the critical level (Ashmore and Wilson, 1994; Fuhrer and Achermann, 1994). For agricultural crops the effects on wheat are used as a reference, and the critical level is defined as the accumulated dose (in ppb hours) above 40 ppb over a period of three months. Only daylight values are used. Two critical levels are suggested; one at 2700 ppb hours representing 5 % loss and one at 5300 ppb hours representing 10 % loss in crop production. Preliminary critical levels are also set for forests.

Inventories of the critical loads/levels and their exceedances are presently being made within the framework of the LRTAP Convention (Downing *et al.*, 1993). The inventories are made with a geographical resolution of better than 150 km. Present inventories show that critical loads and levels are exceeded over large areas in Europe and that far-reaching control measures are necessary to meet requirements for sustainable ecosystems.

For the most recent development of policies for the regional air pollution problems in Europe, *integrated-assessment models* have been developed. With these models the environmental gains can be calculated and optimised in relation to investments in emission control. These models played a central role for the development of the sulfur protocol within the LRTAP Convention signed in 1994. For the development of the strategy of the sulfur protocol, three particular models have been developed, with which scenarios in different sectors can be analysed in terms of the exceedances of critical loads and levels. These models are the RAINS model developed by IIASA, the CSAM model developed by the Stockholm Environment Institute and the ASAM model developed by Imperial College in London. For the control of sulfur emissions in Europe, the models have been used to calculate the most cost-effective control measures in order to reduce the exceedances of critical loads. The results from such a model (RAINS) were used as a basis for the second sulfur protocol signed in June 1994. In all these models the EMEP source-receptor matrices were used to describe the links between emissions and the deposition in the receptor areas. For marine eutrophication no integrated assessment models have yet been developed.

c. Marine eutrophication

Marine eutrophication as an international environmental problem was discovered about 1980 and it was soon realised that atmospheric deposition could be an important source of nutrients. In the marine conventions for the North Sea (PARCOM) and the Baltic Sea (HELCOM), there were agreements on reducing the input of nutrients, including atmospheric deposition, to the sea areas by 50 % by 1995; goals which will not be met (Table 2.5). There have not been any attempts so far to develop effect-oriented cost-effective control strategies of similar types as for the effects handled within the LRTAP Convention. Before effect-based strategies can be negotiated and implemented a number of important scientific issues need to be addressed. These include the role of atmospheric deposition in the total input and effects in marine ecosystems and whether direct deposition of inorganic nitrogen to sea surfaces may give effects in addition to those normally considered, that is the input of mainly organic nitrogen to water (rivers, sewage plants *etc.*).

Table 2.5:   Marine conventions and joint actions in which reductions of regional and global air pollutants may be necessary.

Convention	Area of concern	Atmospheric pollutants of concern	Agreements on reductions
PARCOM	North Sea	Nutrients, POC, some metals	50 % reduction of land based inputs
HELCOM	The Baltic	Nutrients, POC, some metals	50 % reduction of land based inputs
BARCELONA Commission	The Mediterranean		

2.3.2 Policies addressing global environmental problems

Although global environmental problems were not issues of primary concern for EUROTRAC, some of the activities within the project have a bearing on them. Therefore a brief description of the international policy process related to global air pollution problems will be given in this section.

The global air pollution problems, mainly linked to stratospheric ozone depletion and increasing concentrations of greenhouse gases, are of concern to all countries. The natural forum for handling global environmental problems is the UN and its sub-organisations such as UNEP and WMO.

Global environmental concerns and actions are much more complicated to handle in international negotiations since, even if all countries contribute to the emissions, the effects and effect risks may vary substantially from country to country.

Another problem relates to the time scales. Greenhouse gases as well as gases depleting the ozone layer have very long residence times in the atmosphere and the effects of the emissions produced now will last for several decades. Moreover, for the most important greenhouse gas, carbon dioxide, far-reaching emission control requires deep changes in the society. Such changes will take decades and require, not only a common understanding and awareness of the problem, but also costly investments and in many cases technologies not in common use today.

a. Stratospheric ozone

The protection of the stratospheric ozone layer was firstly handled in international bodies in 1976, when a UNEP research programme entitled *Work Plan of Action on the Ozone Layer* was started. The awareness of the threat to the ozone layer required the control of CFCs by a number of countries. It was not however until 10 years later, in connection with the observations on an increased ozone depletion over Antarctica, that an international convention, the *Vienna Convention for the protection of the ozone layer,* was introduced. The Vienna Convention is, as is the LRTAP Convention, a framework Convention which contains no obligations for specific control measures. The obligations appear in the *Montreal Protocol* Vienna Convention for the protection of the ozone layer, signed in 1987. When this protocol was signed, the depletion of the ozone layer over the Antarctica was fully recognised and was the basis for the agreements. Nevertheless it did not require fast cutbacks of the production, and it was quite fair to the developing countries in allowing them access to the controlled substances for long time periods. The protocol has been revised twice; in London 1989 and in Copenhagen in 1992, strengthening the control of the substances agreed earlier and introducing new substances into the protocol. The present status of the Montreal protocol after the Copenhagen amendment and the Vienna Convention for the protection of the ozone layer is specified in the Table 2.6. By late 1992, the Montreal protocol had been signed by more than eighty parties, including all the industrialised countries and more than forty of the developing countries.

Table 2.6:   Control requirements in the Montreal Protocol including the revisions from London 1989 and Copenhagen in 1992.

Chemical	Action
CFCs	Annual production of each CFC must be reduced by 75 % by 1 Jan. 1994 and eliminated by 1 Jan. 1996. (Base years 1986 and 1989).
Halons (Halon 1211, 1301 & 2402)	A total phase-out is required by 1 Jan. 1994.
Carbon tetrachloride	Annual production must be reduced by 85 % by 1 Jan. 1995 and eliminated by 1 Jan. 1996.
Methyl chloroform	Annual production must be reduced by 50 % by 1 Jan. 1994, and by 100% by 1 Jan. 1196.
HCFCs	A freeze in consumption by 1996 and an elimination by 100 % by 2030.
HBFCs	A complete phase out in production and consumption by 1 Jan. 1996
Methyl bromide	A freeze in consumption and production in 1995.

b. Greenhouse gases

The theoretical and experimental evidence of an increase in radiatively active gases resulting in significant changes to the earth's radiation budget, lead to the international *Climate Change Convention* signed by most of the world's countries in Rio de Janeiro 1992, and the *Intergovernmental Panel on Climate Change (IPCC)*. The Panel was jointly established by WMO and UNEP in 1988 in order to assess the scientific information on climate change, together with the environmental and socio-economic impacts, and to formulate response strategies. The Panel has so far published one report (IPCC, 1990) which was updated in 1992 (IPCC, 1992). A second assessment report is expected in 1995.

Besides CO_2, a number of other trace gases have been identified as greenhouse gases. Among them are CH_4, N_2O, a range of halogenated compounds, and O_3.

2.4   EUROTRAC and its Application Project

2.4.1   EUROTRAC: the successful scientific project

a. Foundation and aims

It was with the recognition in the mid-1980s that the scientific problems associated with air pollution could only be solved by an international and inter-disciplinary approach that EUROTRAC was formed. The problems were seen to be beyond the capability of a single laboratory or even a single country to solve; a combined approach was needed.

EUROTRAC is a co-ordinated environmental research programme, within the EUREKA initiative, studying the transport and chemical transformation of trace substances in the troposphere over Europe. It provides a framework in which the resources and research groups from the participating countries can be harnessed to tackle the inter-disciplinary and international transboundary scientific problems directly related to three tropospheric environmental issues which are still of current concern: the formation of photo-oxidants, the production of acidifying substances and eutrophication. It was intended to provide the necessary support for identifying source-receptor relationships for pollutants which in turn are necessary pre-requisites for the development of sound policies for pollutant emission abatement strategies in Europe.

EUROTRAC was accepted as a EUREKA environmental project by the Hannover meeting of Ministers in 1985; work started in 1986 with the definition phase and the implementation phase followed in 1988. The scheduled finish of the project is at the end of 1995. (see EUROTRAC ISS, 1994a for further details).

The stated aims of EUROTRAC are:

*   to increase the basic understanding of atmospheric science;

*   to promote the technological development of sensitive, specific, fast-response instrumentation for environmental research and monitoring; and

*   to improve the scientific basis for taking future political decisions on environmental management within Europe.

b. EUROTRAC: scientific and technological progress

Under the guidance of the Scientific Steering Committee (SSC), the project was organised into fourteen subprojects, each with specific objectives designed to meet the first two aims of EUROTRAC as a whole. Over the intervening years, as the comprehensive annual reports (EUROTRAC ISS, 1990, 1991, 1992, 1993, 1994a), and the proceedings of the Symposia illustrate (Borrell et al., 1991, 1993, 1994), the work done by the principal investigators, within the framework of co-ordination offered by the subprojects, has ensured that the first two aims of EUROTRAC will be achieved.

At the half-way stage of EUROTRAC the International Executive Committee (IEC), which bears the overall responsibility for the project, commissioned a firm of consultants to review the progress within EUROTRAC and to make recommendations for improvements which could be implemented during the second half of the project.

The reviews of seven independent scientific reviewers confirmed that much progress was being made in the scientific and technological programme and that the first two aims were likely to be achieved. However the consultant firm warned that the third aim would not be achieved with the project as then organised since it was directed towards the scientific and technological aims determined, in a "bottom-up" approach, by the scientific groups themselves. A "top-down" approach would be necessary to achieve the third aim of improving the scientific basis for political decision making to determine European air quality. (SERCO Space, 1992)

The conclusion was not unexpected since EUROTRAC was primarily set up as a scientific project. It was driven by the needs for increased scientific knowledge on the behaviour of trace constituents emitted to the atmosphere and their role in influencing the chemistry and properties of the atmosphere. However, while it had the aim of contributing to the development of policy, the project did not include any formal links to policy development, for which other bodies and processes such as the LRTAP Convention already existed (see section 2.3.1). Instead the objective was to establish a scientific edge by which basic knowledge is developed to support the existing policy processes.

Naturally, during the project, a number of contacts and links, formal as well as informal, were established with the LRTAP Convention's body for monitoring pollutants in the atmosphere, EMEP. Formal links on data exchange were established between TOR and EMEP, a group from LACTOZ made an evaluation of the chemistry used for the atmospheric chemistry modules used in the EMEP models, model intercomparisons were carried out with the EURAD model of EUMAC and other models (LOTOS, REM-III), emission data was provided by EMEP for use in the EUROTRAC models and is being provided by EUROTRAC to EMEP for biogenic compounds (see section 3.2.2). Thus whether policy issues were specifically addressed in the projects or not, there are today strong links between science and policy within the project and many of the results from EUROTRAC are already included in the process of application and the development of policy.

However the recommendations of the consultants were clear: a real effort had to be made to convert the wealth of scientific knowledge and expertise accumulated in EUROTRAC into a form suitable for use in policy applications.

2.4.2   The Application Project (AP)

a. The formation and funding of the Application Project

The IEC accepted the principal recommendation of the consultants' report to establish a specific project with the objective of bringing together the scientific results from the work of EUROTRAC and making them available to those responsible for the implementation of environmental policy in Europe. The

Application Project was formed in 1993. It was agreed that the project, unlike all the other work done within EUROTRAC, should be funded centrally by the participating countries and the EC. In addition several countries offered to fund their own participants in the AP directly.

b. The aim of the Application Project

As the project description indicates (Appendix A) the formal aim of the AP is

"to assimilate the scientific results from EUROTRAC and present them in a condensed form, together with recommendations where appropriate, so that they are suitable for use by those responsible for environmental planning and management in Europe".

It was envisaged that the AP would thus contribute directly to fulfilling the third aim of EUROTRAC, that is: "to improve the scientific basis for taking future political decisions on environmental management within Europe".
Whether it has done so can be judged from this report.

c. Scientific and organisational assessment of EUROTRAC: the final scientific report

It should be noted that the AP was *not* expected to report on the scientific progress made by EUROTRAC. This is done in the other volumes of this, the final scientific report of EUROTRAC. Similarly the AP was not intended to assess the success of EUROTRAC as a way of organising research in this area. This was done by the consultants at the half way stage (SERCO Space, 1992) and will be judged on the quality of the final scientific report.

d. The themes addressed by the AP

The AP was requested to address three themes (Appendix A):

i. Photo-oxidants in Europe;
    -   in the free troposphere;
    -   in rural atmospheres;
    -   in urban atmospheres.

ii. Acidification of soil and water and the atmospheric contribution to nutrient inputs;
    -   the variation in type and concentration of acidification as exemplified by the content,
    -   of SO_2, NO_x, NH_3 and organic acids in precipitation should be considered.

iii. The contribution of EUROTRAC to the development of tools for the study of tropospheric pollution, in particular:
    -   tropospheric modelling,

- new or improved instrumentation,
- provision of laboratory data.

The first two themes, photo-oxidants and acidity, reflect the policy orientation of the two main scientific objectives of EUROTRAC:

* The elucidation of the chemistry and transport of ozone and other photo-oxidants in the troposphere.

* The identification of the processes leading to the formation of acidity in the atmosphere, particularly those involving aerosols and clouds.

These were chosen at the outset of EUROTRAC because of the concern throughout Europe about the potential environmental impact of photo-oxidants and acid precipitation.

e. The particular role of the "tools" section

The AP was asked to address the third theme, "the contribution to the development of tools", because it was realised that some of the scientific work done within EUROTRAC has longer term implications for policy development. Such work may not contribute immediately or directly to present policy development but serves to produce, in the widest sense, "tools" which will be used by future investigators whose work itself will contribute directly to policy.

The three "tools" chosen are crucial to the development of our future scientific understanding and to effective environmental policy implementation.

- Simulation models for the troposphere and its chemistry are essential in policy work and yet must be, by their very nature as imperfect representations of current knowledge, under continuous improvement.

- The results from laboratory work contribute directly to assessing the importance of individual reactions in the atmosphere and to improving the understanding of the complex chemical and physical processes taking place in the atmosphere. They find their ultimate use in the improvement of atmospheric models.

- Instrument development is required since our knowledge of air quality stems from instrumental measurements. As our understanding grows, the need for new and better instrumentation is imperative.

Lastly it should be remembered, when considering the relationship of the development of tools to policy development, that the work in EUROTRAC itself is based on the "tools" developed by our scientific predecessors. EUROTRAC would have been incomplete and sterile if it too, while addressing its objectives, had not been developing tools for future use.

f. Ensuring a consensus: the mode of operation of the AP

Following advice from the Application Steering Group (ASG), who had been appointed by the IEC to provide the project with general guidance, a postal survey was initiated inviting the principal investigators and subproject coordinators to provide information on those of their results of direct relevance to policy applications. Of the 250 people contacted some 120 responded. Further information and opinions were sought at a joint meeting of the IEC, SSC and subproject coordinators, and at the EUROTRAC Symposium in Garmisch-Partenkirchen in April 1994.

The AP report was then prepared on the basis of the information provided by the principal investigators, the information in the annual reports of the subprojects and the project as a whole (EUROTRAC ISS, 1990; 1991; 1992; 1993; 1994b), the proceedings of the biennial symposia (Borrell *et al.*, 1991; 1993; 1994) and the publications in the scientific literature.

A draft of the report was circulated throughout EUROTRAC for comment before it was considered formally by the SSC and approved by the IEC in July 1995. Thus every effort was made to produce a report which is truly representative of the views of those involved in the project.

2.4.3   The organisation of the Application Project report

This report is organised like the AP project itself. This introductory chapter sets the scene in terms of the policy background, the development of the various protocols for emission abatement and the necessary accompanying development of the scientific understanding and how the Application Project came into being.

Chapters 3 and 4 then follow on the main topics: photo-oxidants and acidity. Each of these presents the policy issues with the conclusions and recommendations drawn from the EUROTRAC work together with enough scientific background to ensure that the basis for the conclusions and recommendations is clear.

Chapter 5 on "tools" is structured in such a way that it reflects the development of various areas within EUROTRAC which are improving the basis for the application of science in policy development in the future.

The report finishes with a series of appendices giving necessary information about the Application Project, the subprojects and EUROTRAC itself.

2.5   Acknowledgements

The members of the Application Project would like gratefully to acknowledge the help and encouragement they have received from the following people or groups.

- The principal investigators and subproject coordinators of EUROTRAC.

- The members of the Application Steering Group,

    IEC chairman and vice chairman: Erik Fellenius and Gerhard Hahn

    SSC chairman and SSC member: Anthony R. Marsh and Dieter Kley

    Director of the International Scientific Secretariat: Wolfgang Seiler

- The members of the International Executive Committee and Scientific Steering Committee (SSC)

- A number of other colleagues in the EMEP centres, the EC and elsewhere who have kindly provided information and comment.

- The following countries which funded the Application Project through their contributions to the support of the ISS.

Austria	Netherlands
Belgium	Norway
Denmark	Sweden
Finland	Switzerland
France	Turkey
Germany	United Kingdom
Greece	

together with the European Union

- The following countries which also supported directly participants from their own country in the Application Project (AP)

Netherlands	Sweden
Norway	United Kingdom

2.6   References

Andersson, L. and Rydberg, L., 1988, Trends in nutrient and oxygen conditions within the Kattegat: effects of local nutrient supply. *Estuarine, Coastal and Shelf Science* **26**, 559–579.

Ashmore, M.R. and Wilson, R.B.,1994, Critical Levels of Air Pollutants for Europe. *Background Papers for the UN ECE Workshop on Critical Levels*, Egham UK 23–26 March 1992. UK Department of Environment Report, London.

Ashmore, M.R., Bell, J.N.B. and Brown, I.J., 1990, Air pollution and forest ecosystems in the European Community. *Air Pollution Research Report* **29**, Brussels.

Atkins, D.H.F., Cox, R.A. and Eggleton, A.E.J., 1972, Photochemical ozone and sulfuric acid formation in the atmosphere over southern England. *Nature* **235**, 372–376.

Berg, J. and Radach, G., 1985, Trends in nutrient and phytoplankton concentrations at Helgoland Reede (German Bight) since 1962. *ICES, C.M. 1985/L:2/Sess. R.* ICES Copenhagen, Denmark.

Bobbink, R., D. Boxman, E. Fremstad, G. Heil, A. Houdijk and J. Roelofs, 1992, Critical loads for nitrogen eutrophication of terrestrial ecosystems based upon changes in vegetation and fauna. in: Grennfelt P. and Thörnelöf, E. (eds) *Critical Loads for Nitrogen. Report from a workshop held at Lökeberg, Sweden 6–10 April 1992. Nord 1992:***41,** Nordic Council of Ministers, Copenhagen, pp. 111–160.

Borrell, P., Borrell P.M. and Seiler, W. (eds), 1991, Transport and transformation of pollutants in the troposphere, *Proc. EUROTRAC Symp. '90*, SPB Academic Publishing, The Hague 1991, pp. xxiii + 586.

Borrell, P.M., Borrell, P., Cvitaš, T. and Seiler, W. (eds), 1993, Photo-oxidants, precursors and products, *Proc. EUROTRAC Symp. '92*, SPB Academic Publishing bv, The Hague 1993, pp xxxi + 832.

Borrell, P.M., Borrell, P., Cvitas, T., and Seiler, W. (eds), 1994, Transport and transformation of pollutants in the troposphere, *Proc. EUROTRAC Symp. '94*, SPB Academic Publishing bv, The Hague 1994, pp xxxi + 1295.

Bouscaren, R., 1991, The problems related with the photochemical pollution in the Southern E.C. member states, *Final Report,* Contract No **6611–31–89**.

CEC, 1988, Directive limiting emissions of atmospheric pollutants from large combustion plants. *Official J. of the European Community* **L 336,** 07.12.88.

Charlson, R.J., Langner, J., Rodhe, H., Leovy, C.B. and Warren, S.G., 1991, Perturbation of the Northern hemispheric radiative balance by backscattering from anthropogenic sulfate aerosols. *Tellus* **43A(13)** 152–163.

Cox, R.A., Eggleton, A.E.J., Derwent, R.G., Lovelock, J.E. and Pack, D.H., 1975, Long range transport of photochemical ozone in north-western Europe. *Nature* **255**, 118–121.

Downing, R.J, Hettelingh, J-P. and de Smet, P.A..M., 1993, Calculations and Mapping of Critical Loads in Europe: *Status Report for the UN-ECE Convention on Long-range Transboundary Air Pollution 1993*. RIVM, Bilthoven, The Netherlands

Eliassen, A., Hov, Ø., Isaksen, I.S.A. Saltbones, J. and Stordal, F., 1982, A Lagrangian long-range transport model with atmospheric boundary layer chemistry. *J. App. Meteorology* **21**, 1645–1661.

EUROTRAC ISS, 1990, *EUROTRAC Annual Report for 1989, parts 1 to 9,* EUROTRAC ISS, Garmisch-Partenkirchen.

EUROTRAC ISS, 1991, *EUROTRAC Annual Report for 1990, parts 1 to 9,* EUROTRAC ISS, Garmisch-Partenkirchen.

EUROTRAC ISS, 1992, *EUROTRAC Annual Report for 1991, parts 1 to 9,* EUROTRAC ISS, Garmisch-Partenkirchen.

EUROTRAC ISS, 1993, *EUROTRAC Annual Report for 1992, parts 1 to 9,* EUROTRAC ISS, Garmisch-Partenkirchen.

EUROTRAC ISS, 1994a, *EUROTRAC Brochure*, EUROTRAC ISS, Garmisch-Partenkirchen, pp 20.

EUROTRAC ISS, 1994b, *EUROTRAC Annual Report for 1993, parts 1 to 9,* EUROTRAC ISS, Garmisch-Partenkirchen.

Fuhrer, J. and Achermann, B., 1994, Critical Levels for Ozone; *Report from a UN-ECE workshop held at Bern, Switzerland 1–4 November 1993*, Federal Research Station for Agricultural Chemistry and Environmental Hygiene, Liebefeld-Bern, Switzerland.

Grennfelt, P. and Thörnelöf, E. (Eds.), 1992, Critical Loads for Nitrogen. *Report from a workshop held at Lökeberg, Sweden, 6–10 April 1992.* Nordic Council of Ministers, Copenhagen.

Grennfelt, P., Hov, Ø. and Derwent, R.G., 1994, Second generation abatement strategies for NO_x, NH_3, SO_2 and VOC. *Ambio* **23**, 425–433.

Guderian, R. and Tingey, D. T., 1987, Notwendigkeit und Ableitung von Grenzwerten für Stickstoffoxide. *UBA Berichte* **1/87.** Umweltbundesamt, Berlin.

Gundersen, P., 1992, Mass balance approaches for establishing critical loads for nitrogen in terrestrial ecosystems. In Grennfelt P. and Thörnelöf, E. (Eds.) Critical Loads for Nitrogen. *Report from a workshop held at Lökeberg, Sweden 6–10 April 1992. Nord 1992*:**41**, Nordic Council of Ministers, Copenhagen, pp. 55–109.

Haagen-Smit, A.J., 1952, Chemistry and physiology of Los Angeles smog, *Ind. Eng. Chem.* **44**, 1342–1346.

Hallbäcken, L. and Tamm, C.O., 1986, Changes in soil acidity from 1927 to 1982–84 in a forest area in south-west Sweden. *Scandinavian J. Forest Res.* **1**, 219–32.

Hov, Ø., Hesstvedt, E. and Isaksen, I.S.A., 1978, Long-range transport of tropospheric ozone. *Nature* **242**, 341–344.

Hultberg, H. and Stensson, J., 1970, The effects of acidification on the fish fauna in two lakes in the county of Bohuslän, *Sweden Fauna & Flora* **1**, 11–19 (In Swedish).

Innes, J.L., 1993, *Forest Health: Its Assessment and Status.* CAB International, UK.

IPCC, 1990, Houghton, J.T., G.J. Jenkins, and J.J. Ephramus, Climate Change, *The IPPC Scientific Assessment, Intergovernmental Panel on Climate Change*, Cambridge University Press, New York.

IPCC, 1992, Climate Change 1992, *The IPCC Scientific Assessment*, Intergovernmental Panel on Climate Change, Cambridge University Press.

Jensen, K. and Snekvik, E., 1972, Low pH levels wipe out salmon and trout populations in southern most Norway. *Ambio* **1**, 223–225.

Nilsson, J. and Grennfelt, P. (eds), 1988, Critical Loads for Sulfur and Nitrogen. Critical Loads for Nitrogen. *Report from a workshop held at Stockholm, Sweden, 19–24 March 1988.* NORD miljørapport 1988: **15**, Nordic Council of Ministers, Copenhagen.

Odén, S.,1968,. The acidification of air precipitation and its consequences in the natural environment. *Ecological Committee Bulletin* **1**, Stockholm.

OECD, 1977, THE OECD programme on long range transport of air pollutants, measurements and findings. Organisation for Economic Co-operation and Development, Paris 1977.

SERCO 1992, *Review of EUROTRAC*, SERCO SPACE Ltd., Sunbury-on Thames, pp 160.

Ulrich, B. , Matzner, E., 1983, Abiotsche Folgewirkungen der Weitraumigen Ausbreitung von Luftverunreinigungen, Forschungsbericht 104 02 615.

UN-ECE, 1994, Protocol to the 1979 Convention on Long-Range *Transboundary Air Pollution on Further Reduction of Sulfur Emissions.* ECE/EB.AIR/40. Geneva.

UN-ECE, 1988, UN-ECE critical levels workshop report, Bad Harzburg 14-18 March 1988, Umweltbundesamt, Germany.

UN, 1987, *Our Common Future.* Oxford University Press.

Chapter 3

Photo-oxidants

3.1 Environmental policy issues related to photo-oxidants

In the development of control policies for photo-oxidants, questions must be addressed which relate to the damage inflicted by elevated concentrations on human health, plants and crop yields, how atmospheric properties are affected, the geographical area which is influenced and the change in severity with time. The link between man-made emissions and elevated concentrations must be established, and the emission reductions as well as costs involved to meet specific environmental targets, need to be determined.

The most important photochemical oxidants are ozone, O_3, nitrogen dioxide, NO_2, peroxyacetyl nitrate, PAN, and hydrogen peroxide, H_2O_2. The effects of elevated concentrations are linked both to peak values in episodes of limited duration, and to a general increase in the background concentrations.

In urban areas the impact of photochemical oxidants on human health in episodes is of particular significance, with exposure to high concentrations of O_3, NO_2 and possibly other compounds derived from emissions of volatile organic compounds and oxides of nitrogen in urban areas (nitrous acid, nitrate radicals and PAN).

Over industrialised and populated continents, the impact of elevated concentrations of photo-oxidants on health, agriculture, forests and natural ecosystems and materials is linked both to peak values in episodes of one or a few days' duration, and to concentrations above critical levels averaged over the daylight hours in a growing season. The concept of a critical level for ozone assumes a threshold, below which no or small effects occur. The observed effects are then related to the ozone dose expressed in ppb-hours above the threshold levels. A threshold of 40 ppb for ozone is proposed by the UN-ECE Task Force on Mapping for daytime sampling, with no effects if an ozone dose of 300 ppb-hours is not exceeded. This critical level is exceeded every year all over Europe, with the highest exceedances being in central Europe.

Changes in the concentration of tropospheric ozone have an impact on the oxidising capacity of the atmosphere, on climate, on the ground level concentration of ozone, and on the penetration of ultraviolet radiation to the earth's surface. It is of particular importance to determine the size and causes of the upward trend seen in tropospheric ozone over Europe, and in the northern hemisphere in general, over the recent decades.

In this chapter policy-oriented questions will be addressed, linked to the quantification of emissions, their transformation into products, their dispersion and removal, and the integration of processes into models in order to establish source-receptor relationships and construct the budgets for specific species of photo-oxidants. Questions about how elevated levels of photo-oxidants affect health, ecosystems and materials and about costs and time required to implement specific policies, will not be discussed since EUROTRAC was not set up to provide such information.

The following general policy-oriented questions will be addressed:

- To what extent do human activities contribute to the observed distributions and trends with time of photo-oxidant concentrations over Europe? Are there important biogenic contributions as well?

- Which emission reductions are required to fulfil specific environmental policies?

The discussion of photo-oxidants is separated into three different parts characterised by their dimensions in time and space, as shown in Table 3.1

Table 3.1:   Scales for consideration of photo-oxidants

	vertical scale	horizontal scale	time scale	compounds
Local or urban	2 km	100 km	1 day	O_3, NO_2, PAN, HONO
Regional or continental	2 km	2500 km	5–100 days *	O_3, PAN, H_2O_2, NO_2
Global	10 km	20000 km	100 days	O_3, PAN, NO_2

* 100 day time scale is linked to the growing season average ozone concentration.

3.2   Scientific findings of importance for environmental policy

To find out how EUROTRAC has contributed or can contribute to answering policy-oriented questions, the relevant scientific findings in EUROTRAC as well as in other research will now be assessed. Such scientific findings are linked to the understanding of mechanisms and processes, for example in the quantification of emissions, their transformation into products, dispersion and removal, and the integration of processes into models in order to establish source-receptor relationships and construct the budgets for specific species.

3.2.1   Trends in photo-oxidant concentrations and precursors

In trend studies an attempt is made to establish what the concentrations of photo-oxidants would have been without the intervention of man. Analyses of available records, for example those from Hohenpeissenberg (Attmannspacher *et al.*, 1984; Wege *et al.*, 1989) and Arkona (Feister and Warmbt, 1987), and the re-analysis of historic measurements (Volz and Kley, 1988; Kley *et al.*, 1988; Crutzen, 1988) have suggested that the ozone concentrations in the 1980s were higher than what they were around the turn of the century. Photochemical formation is the likely cause for that increase. Regional model simulations also clearly indicate that there would be lower ozone levels in the unpolluted boundary layer over Europe, if anthropogenic emissions were excluded (Ebel *et al.*, 1991). A clearer picture has now emerged of the magnitude and the nature of the increase in tropospheric ozone concentrations.

a. Evaluation of historical ozone records

Widespread measurements of ozone were made in the last century, mostly using the Schönbein test paper, which is subject to interference from wind speed (Fox, 1873) and humidity (Linvill *et al.*, 1980). Kley *et al.* (1988) concluded from an extensive laboratory evaluation of the method that these data are only semi-quantitative in nature and should not be used for trend estimates. A quantitative method to measure ozone was used continuously from 1876 until 1911 at the Observatoire de Montsouris, Paris (Volz and Kley, 1988). The 24 h average ozone concentration was around 10 ppb, about a factor of 3 to 4 smaller than that found today in many areas of Europe and North America.

Analysis of further Schönbein records led to similar conclusions about the pre-industrial ozone concentration and would suggest that the tropospheric background was 10 ppb in both hemispheres (see Anfossi *et al.*, 1991; Sandroni *et al.*, 1992) and in the free troposphere over Europe (Marenco *et al.*, 1994). This agreement must be viewed with caution however, because of the known problems associated with the Schönbein method. In particular, the close agreement between ozone concentrations at Pic du Midi at 3000 m altitude and at Montsouris in the Paris basin is difficult to explain because, in the absence of local photochemical production, daily average ozone concentrations at a surface site like Paris, which is heavily influenced by dry deposition at night, are expected to be lower than those in the free troposphere.

After 1910, only few mostly sporadic measurements of tropospheric ozone were made using both optical and chemical techniques, reviewed by Crutzen (1988), Kley *et al.* (1988) and Staehelin *et al.* (1994).

In Fig. 3.1 a comparison of historic measurements that were made using quantitative techniques is shown with measurements made in the late 1980s. On average, ozone concentrations in the troposphere over Europe (0 to 4 km) today are a factor of two higher than in the earlier period. Fig. 3.1 also shows that little can be inferred about a possible increase in tropospheric ozone before 1950, because of the variance between the different sites. In this context, it is interesting to note that the TOR data from Montsouris (1876–1911; 40 m a.s.l.) and those from Arosa (1950-1956; 1860 m a.s.l.) do not show a single day with ozone concentrations above 40 ppb (Volz-Thomas *et al.*, 1993; Staehelin *et al.*, 1994).

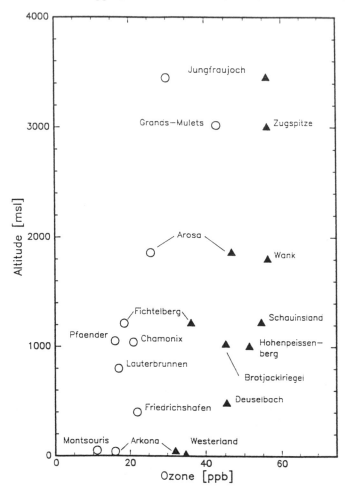

Fig. 3.1:   Surface ozone concentrations observed in late summer at different locations in Europe (based upon Staehelin *et al.*, 1994). The open circles summarise historic data collected before and during the 1950s, the triangles are from measurements made after 1988. The data have been plotted against the altitude of the different sites.

b. Trends in ozone concentrations over the past decades

The modern ozone measurements are mainly based on UV absorption, and were started in the 1970s at several remote coastal and high altitude sites. The records for Mauna Loa, Hawaii (Oltmans and Levy, 1994) and the Zugspitze, southern Germany (Scheel *et al.*, 1993) are shown in Fig. 3.2.

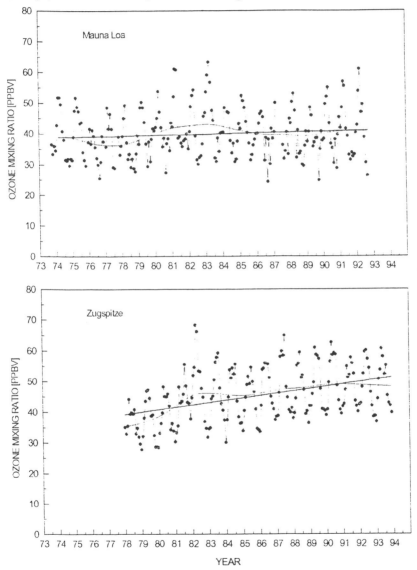

Fig. 3.2:   Trends in tropospheric ozone concentrations observed at the TOR-station Zugspitze since 1978 (based upon Scheel *et al.*, 1993; Sladkovic *et al.*, 1993) in comparison to the trend observed at Mauna Loa, Hawaii since 1973 (based upon Oltmans and Levy, 1994).

The trends observed at the various remote sites were evaluated in TOR and are presented in Fig. 3.3. All stations north of about 20 °N exhibit a positive trend in ozone that is statistically significant. On the other hand, a statistically significant negative trend of about 7 % per decade is observed at the South Pole. For the most part, the trends increase from –7 % per decade at 90 °S to +7 % per decade at 70 °N. There are particularly large positive trends observed at the high elevation sites in southern Germany (1–2 % per year). These large trends perhaps reflect a regional influence above and beyond the smaller global trend (Volz-Thomas, 1993).

The trends observed in the northern hemisphere are largely due to the relatively rapid ozone increase that occurred in the 1970s. Over the last decade, no or only a small ozone increase has occurred in the free troposphere and, indeed, ozone concentrations at some locations in the polluted boundary layer over Europe have even decreased during the last decade. At Delft in the Netherlands ground level ozone concentrations decreased in the 1970s as a result of increasing NO_x concentrations, with $O_x = O_3 + NO_2$ increasing slightly (Guicherit, 1988; Low et al., 1992). In the 1980s and 1990s ground level ozone concentrations as well as the oxidant (O_x) levels, as measured in Kollumerwaard in the northern part of the Netherlands, decreased. Measurements of ozone at Mace Head in the period from April 1987 to June 1992 show an average upward trend of 2.5 ppb per decade in

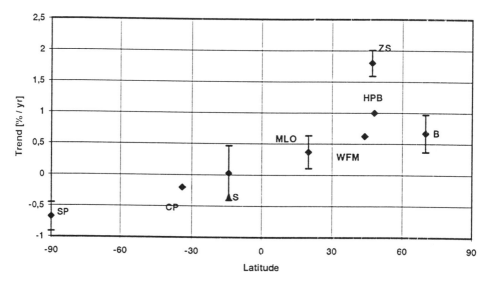

Fig. 3.3:   Trends in tropospheric ozone concentrations observed at different latitudes. Only coastal and high altitude observatories are included. SP: South Pole, 90 °S, 2800 m a.s.l., start 1975; CP: Cape Point, 34 °S, 1982; S: Samoa, 14 °S, 1975; MLO: Mauna Loa, 20 °N, 3400 m a.s.l., 1973; WFM: Whiteface Mountain, 43 °N, 1600 m a.s.l., 1973; ZS: Zugspitze, 47 °N, 3000 m a.s.l., 1978; HPB: Hohenpeissenberg, 48 °N, 1000 m a.s.l., 1971; B: Barrow, 70 °N, 1973. (based upon Volz-Thomas, 1993).

the summer in polluted air which has passed over continental Europe. In polluted winter air there is a downward trend of 1.3 ppb per decade (Simmonds, 1993). It will be possible to derive more reliable estimates for the trends when a longer time series is available.

Ground based instruments do not often sample free tropospheric air, but at marine boundary layer sites like those at Samoa and Cape Point in South Africa Cape Point, South Africa the concentration trends derived are representative of large geographical regions.

Free tropospheric concentrations are obtained from ozone sondes or from measurements at high altitude sites such as the South Pole, Mauna Loa and the Zugspitze. London and Liu (1992), Logan (1994) and Miller *et al.* (1994) have analysed the global ozone sonde records. All the studies of the ozone sonde records show, on average, increases in free tropospheric ozone at northern mid-latitudes of around 10 % per decade since 1970. Although the North American records are not of the same length and quality as those from Europe, it seems likely that the trend in free tropospheric ozone over North America has been smaller than that observed over Europe. New studies also indicate that the upward trend over Europe is smaller since 1980 than earlier.

Logan (1994) argues that the measurements made at Hohenpeissenberg, Lindenberg and possibly other European stations might be influenced by SO_2. In polluted areas, local titration of ozone by NO_x can also influence measurements of ozone, but these effects should not be important out of the atmospheric boundary layer. De Muer and De Backer (1994) have corrected the Uccle data set allowing for the known instrumental effects, including SO_2 interference. The ozone trend in the upper troposphere was only slightly reduced (10–15 % per decade, 1969–91) and remained statistically significant. However, below 5 km the trend was reduced from around +20 % per decade to +10 % per decade and became statistically insignificant.

The ozone sonde records from three Japanese stations between 1969 and 1990 (Akimoto *et al.*, 1994) indicate annual trends of 25 ± 5 % per decade for the 0 to 2 km layers, and 12 ± 3 % per decade for the 2 to 5 km layers. Between 5 and 10 km, the trend is 5 ± 6 % per decade. There is no evidence for a smaller trend in the 1980s. In the tropics, Logan (1994) reports that in Natal there has been an increase in ozone between 400 and 700 hPa. The Melbourne record shows that the only significant decrease in tropospheric ozone is between 600 and 800 hPa and is largest in summer.

c. Trends in precursor concentrations

Direct measurements are not available for the analysis of the long-term trend in NO_x and hydrocarbon concentrations. The longest continuous record for individual hydrocarbons is from Birkenes near the south coast of Norway, and since the summer of 1987 a part of the subproject TOR. Solberg *et al.* (1994b) have shown that there is a statistically significant upward trend in the concentrations of acetylene, propane, butane and also in the sum of C_2–C_5 hydrocarbons. On the other hand, there is a downward trend in the concentration of alkenes (ethene and propene). By comparing the observed changes in nonmethane hydrocarbon concentrations with changes in the large scale transport patterns, Solberg *et al.* (1994b) concluded that, in addition to changes in emissions, climatological variability is an important factor for the observed changes. A clear picture of how emission reductions affect the atmospheric concentrations of fairly short-lived species will only be possible with the help of longer measurement series at more locations.

Indirect information on long-term trends in NO_x can be obtained from the analysis of ice cores. Fig. 3.4 shows the most recent analysis from subproject ALPTRAC, of the concentration of nitrate and lead in an ice core from a high-altitude Alpine glacier (Wagenbach *et al.*, 1993, Schajor *et al.*, 1994). The concentrations of both species show strong increases after 1940, after having remained almost constant in the period before. Nitrate is the final product of NO_x oxidation and is removed from the atmosphere by heterogeneous processes such as rain-out. Therefore changes in the concentration of nitrate in the ice shouldreflect the changes in the concentrations of the precursor NO_x. The onset of the increase in nitrate concentrations coincides with the start of the increase in tropospheric ozone over Europe between 1940 and 1950 (Kley *et al.*, 1988 and Staehelin *et al.*, 1994). Furthermore, the simultaneous increase in the lead and nitrate concentrations indicate that automobile exhaust is a major source of nitrate at Monte Rosa and hence a major cause of the ozone increase.

d. Trends in other photo-oxidant concentrations

Long-term records of other photo-oxidants, such as hydrogen peroxide or peroxyacetyl nitrate, are sparse. For peroxyacetyl nitrate, a continuous record from the Dutch air quality network shows an increase of almost a factor of 3 in the 1970s (Guicherit, 1988) when O_x concentrations were increasing slightly and ozone concentrations decreasing (see above). PAN is formed from peroxy radicals and NO_2 whereas ozone is formed from the reaction of peroxy radicals with NO. The PAN increase suggests that the potential for the formation of photo-oxidants was still increasing in the 1970s in the more heavily polluted areas of Europe. In the 1980s the concentration of PAN at Delft stabilised; a similar result was seen in the 1990s at Kollumerwaard.

The atmospheric measurements of hydrogen peroxide made in the US and in Europe, many of the latter made in the framework of TOR, are not of a sufficient length to allow a trend assessment; the concentration of H_2O_2 is also highly variable in space and with time. Sigg and Neftel (1991) used a record derived from Greenland ice cores to argue in favour of an increase in atmospheric hydrogen peroxide concentrations, but the integrity of such a reactive and light sensitive species in the firn layer, *i.e.* before the ice is formed, needs to be established.

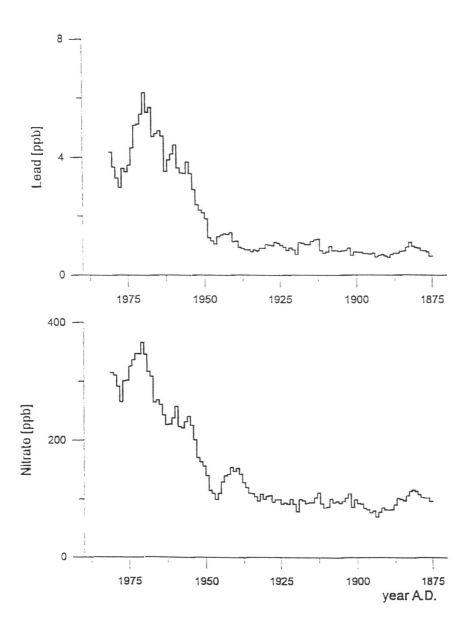

Fig. 3.4:   Long-term trends in lead and nitrate concentrations in the free troposphere over western Europe as derived from a high Alpine ice core study (at the Monte Rosa, 4500 m a.s.l. Wagenbach, unpublished data).

3.2.2   Emissions

A quantification of emissions of the oxides of nitrogen and volatile organic compounds is required to establish source-receptor relationships and construct budgets for specific species. The quantification is required both in terms of source categories (biogenic, industrial, transportation *etc.*), geographical distribution of sources and the variation of emissions with time.

Emission data for nitrogen oxides and VOCs have been collected for many years over Europe including the former Soviet Union as far east as the Urals, within the ECE-EMEP framework, and later in the Dutch-German PHOXA project which was also supported by OECD and the CEC (Builtjes, 1985; Pankrath, 1989). The data for individual grid squares (150×150 km^2) and for the countries themselves often show considerable differences between the inventories.

Hourly data on anthropogenic emissions of SO_2, NO_x, VOC, NH_3 and CO are now provided for several parts of Europe calculated for the most important economic sectors and for area and point sources. In GENEMIS, data have been established for the seasonal variation of all the anthropogenic emission source categories, in particular for power plants and road traffic. For studies on the urban and local scale, the sources of photo-oxidant precursors need to be known with high resolution. Based on CORINAIR 1985 and 1990, LOTOS 1985, 1986 and 1990 and other national and international annual inventories, hourly emission data have been provided for the modelling subproject EUMAC in the EURAD grid for episodes between 1985 and 1994.

GENEMIS has further contributed considerably to harmonisation and improvement of the EU CORINAIR-85 data base. National inventories and emission data have been provided by experts from several countries, including non-EU member states. GENEMIS has also contributed to the work of the UN-ECE Task Force for Emission Inventories (see Heymann, 1993). A system has been developed in Austria to generate time resolved emissions for CO, NO_x and VOC where different technological and legislative options have been introduced (Winiwarter *et al.*, 1994).

Work in the subproject EUMAC has shown that shortcomings in emission data can be detected when observations and model calculations based on accurate meteorological input are compared (Ebel *et al.*, 1994). Application of the EURAD model in this way with EMEP precursor emission data has pointed to an appreciable underestimation of the NO_x emissions in Poland and the former Soviet Union (Ebel *et al.*, 1993).

Results from GENEMIS show that there is a large variation of emissions in Europe with time, and that it differs considerably from country to country. For example, the fuel consumption in power plants in Germany, Italy and the UK shows a regular seasonal variation between 30 and 40 %. However in France it is much larger. From May to August, the high share of nuclear power means that the fuel consumption is almost zero, but in February there is a peak almost three times the average (Friedrich *et al.*, 1994; Hansen *et al.*, 1994). For road traffic the difference in emissions between night and day is usually more than 90 %; during the weekend anthropogenic emissions are about 30 % lower than during the week. The VOC/NO$_x$ ratio (anthropogenic emissions only) is much higher within the cities than outside them. Variations of this magnitude have not been fully accounted for in model calculations of atmospheric chemistry and dispersion, and may help to explain discrepancies between measured and calculated concentrations of photo-oxidants in the vicinity of strong sources. There is some evidence in Belgium that a weekend reduction in emissions leads to an increase in ozone levels (Dumont *et al.*, 1995). The information is also important for smog-alert systems and for the effectiveness of abatement measures during episodes. Time functions and time factors have been prepared for emissions from the most important economical sectors in the different European countries, including specification of the composition of the VOC emissions from different sectors. Some experiments to support and validate emission measurements have been undertaken, for instance in a road tunnel in Switzerland (Staehelin and Schläpfer, 1994). A comprehensive speciation of anthropogenic VOC emissions has been worked out for the United Kingdom with more than 70 species distributed in 8 source categories (Derwent and Jenkin, 1990).

Biogenic emissions have been quantified for Europe by Lübkert and Schöpp (1989) and Veldt (1989) and they have been improved within BIATEX. In GENEMIS the land-use data for Europe have been improved (Kible and Smiatek, 1994), and leaf area index data have been derived from satellite radiometer observations. Forests cover perhaps 40 % of the European land area and are important emitters of isoprene, terpenes as well as oxygenated and other VOCs. It is difficult to derive general data about biogenic emissions since their intensity and composition depend on many factors such as vegetation cover and type, physiological conditions, wind, temperature and insolation (Fehsenfeld *et al.*, 1992; Haymann, 1994; Steinbrecher *et al.*, 1994; Steinbrecher and Rabong, 1994). Emissions should primarily be related to the physiological activity of the plant (Slanina *et al.*, 1994). Helas (1994) summarised the work in BIATEX up to 1993 and concluded that the information presently available is not sufficient to derive a general inventory. Nevertheless, generalisations have been made by Hewitt and Street (1992) for the UK; Simpson (1994a, 1994b) estimated the biogenic VOC emissions in Europe for EMEP modelling studies and Fehsenfeld *et al.* (1992) reviewed global emission estimates. A biogenic emission inventory was derived for EURAD from the work of Lübkert and Schöpp (1989). Several studies within TOR indicate that biogenic hydrocarbon emissions can affect the oxidant formation in rural areas considerably (Klemp *et al.*, 1993; Kramp *et al.*, 1994; Solberg *et al.*, 1994b; Poppe *et al.*, 1994).

In an assessment of the European biogenic VOC emissions for 1989 Simpson (1994b) found that the isoprene emissions were about 6 % of the man-made VOC emissions. If other biogenic emissions besides isoprene were added as well, the annual emissions were estimated at about 38 % of the man-made emissions. The uncertainties are however large. For some countries in Europe (notably in southern Europe) the biogenic VOC emissions are thought to exceed the man-made sources in the summer. The biogenic emissions will have a marked influence on the photo-oxidant formation in summer and consequently on the source-receptor relationship for ozone and other secondary products. Based on model calculations, Simpson (1994b) found that both the time, location and magnitude of the episodic peaks in ozone concentrations change considerably when the isoprene emissions are varied within their range of uncertainty (a factor of 5). However the long-term averages over Europe are not much affected by the uncertainty in the isoprene emissions.

3.2.3   Processes influencing the concentration of photo-oxidants

Photo-oxidants are formed when oxides of nitrogen and volatile organic compounds react under the influence of sunlight. The atmospheric lifetime of the secondary products is often much longer than for the precursors, which means that they have more time to be transported before removal than many of the precursors. A realistic description of the conversion processes and the removal through dry deposition to vegetation is important for the calculation of the concentration of photo-oxidants as well as their source-receptor relationships.

The production of ozone in the troposphere takes place in a complex series of reactions of which the main features have been known since the early 1950s. The role of the hydroxyl radical was established in the early 1970s. Ozone is formed through the photo-oxidation of volatile organic compounds and CO in the presence of NO_x ($NO_x = NO + NO_2$). Typical of this mechanism are reactions (R1) to (R7).

$$RH + OH \rightarrow R + H_2O \tag{R1}$$

$$R + O_2 + M \rightarrow RO_2 + M \tag{R2}$$

$$RO_2 + NO \rightarrow RO + NO_2 \tag{R3a}$$

$$RO + O_2 \rightarrow HO_2 + R'CHO \tag{R4}$$

$$HO_2 + NO \rightarrow OH + NO_2 \tag{R5}$$

$$NO_2 + h\nu \rightarrow NO + O \qquad 2\times(R6)$$

$$O + O_2 + M \rightarrow O_3 + M \qquad 2\times(R7)$$

$$\overline{RH + 4\,O_2 + 2\,h\nu \rightarrow R'CHO + H_2O + 2\,O_3} \qquad \text{net}$$

Thus the initial reaction between a hydrocarbon, RH, and an OH radical results in the production of two O_3 molecules and an aldehyde R'CHO or a ketone. Additional ozone molecules can then be produced from the degradation of R'CHO. Ozone can also be generated from CO oxidation via (R8) and (R9) followed by (R5), (R6), and (R7).

$$CO + OH \quad \rightarrow \quad CO_2 + H \tag{R8}$$

$$H + O_2 + M \quad \rightarrow \quad HO_2 + M \tag{R9}$$

The *in situ* rate of formation of ozone is given by

$$P(O_3) = k_5 [NO] [HO_2] + \Sigma k_{3ai} [NO] [RO_2]_i$$

Photochemical loss of tropospheric ozone is accomplished through photolysis followed by reaction of the $O(^1D)$ atom with water vapour in reactions (R10) and (R11). Additional losses occur through reaction of the HO_2 radical formed in (R9) with O_3 via (R12) and, to a lesser extent, through reaction of OH with O_3 (R13).

$$O_3 + h\nu \quad \rightarrow \quad O(^1D) + O_2 \tag{R10}$$

$$O(^1D) + H_2O \quad \rightarrow \quad 2\ OH \tag{R11}$$

$$HO_2 + O_3 \quad \rightarrow \quad OH + 2\ O_2 \tag{R12}$$

$$OH + O_3 \quad \rightarrow \quad HO_2 + O_2 \tag{R13}$$

Hydrocarbons and CO provide the fuel for the production of tropospheric ozone and are consumed in the process. NO_x is conserved in the process of ozone production and thus acts as a catalyst in ozone formation. The conversion of NO to NO_2 by peroxy radicals (HO_2 and RO_2) is the rate determining step. NO_x acts as a catalyst as long as it is not permanently removed through deposition or rainout, or transformed to other NO_y compounds ($NO_y = NO_x$ + compounds derived from NO_x) which act as temporary or almost permanent reservoirs. The catalytic production efficiency of NO_x can be defined as the chain length, that is the number of NO $\rightarrow$ NO_2 conversions via NO + HO_2 (or RO_2) reactions per NO_2 molecule before NO_x is removed, for example, to nitrate:

Chain length = $(k_5[NO][HO_2] + \Sigma k_{3ai}[NO][RO_{2i}])/(k_1[NO_2][OH] + k_{15}[NO_2][O_3])$

where

$$NO_2 + OH + M \quad \rightarrow \quad HNO_3 + M \tag{R14}$$

$$NO_2 + O_3 \quad \rightarrow \quad NO_3 + O_2 \tag{R15}$$

(R15) acts as a loss process for NO_2 at night. The kinetics of the primary oxidation steps of many of the hydrocarbons were quite well known by the mid 1980s, but there were gaps and uncertainties that prevented the evaluation of the relative role of individual hydrocarbons in oxidant formation, including the lack of speciated emission data for the VOCs. The chemistry of secondary processes which determine the production and loss of radicals was less well known.

The knowledge of the chemistry related to the tropospheric ozone formation has improved considerably through EUROTRAC research (see also section 5.3). The impact of the aromatics on ozone formation can now be assessed quantitatively (Becker, 1994), the ozonolysis of alkenes has been shown to be an important sink for ozone and a source of radicals (Gaeb *et al.*, 1994) and heterogeneous processes have been shown to affect significantly oxidation both in the urban and in the clean atmosphere. The role of radical-radical reactions and chemistry related to the night-time chemistry of NO_y which proceeds via formation of NO_3 radicals, has been investigated in LACTOZ (Wayne *et al.*, 1991). These findings have improved the understanding of chemical reaction mechanisms.

Oxidants other than OH may be of importance in the troposphere; among these are chlorine atoms (Pszenny *et al.*, 1993), which may be formed in the marine boundary layer from reactions of N_2O_5 and NO_2 with aerosol chloride and bromide (Finlayson-Pitts *et al.*, 1989; Zetsch and Behnke, 1992a,b) and NO_3 radicals. NO_3 radicals are formed at night from the oxidation of NO_2 by O_3 and may play a significant role in the atmospheric oxidation of nonmethane hydrocarbons, particularly at high and mid-latitudes (Penkett *et al.*, 1993).

3.2.4   Local and urban photo-oxidants

Photochemical air pollution was first thought to affect primarily urban areas. For this reason, the first studies dealt with the situation in large cities, both in the US and in Europe. Urban emissions were found to be the source of photo-oxidants in the Los Angeles smog in the 1950s (Haagen-Smit, 1952), and in the 1970s it was found that photochemical episodes can be associated with long range transport of ozone and its precursors. It was also recognised that this is more often the case in northern and western Europe than in the south of the continent, where photo-smog is often of a local character (Bouscaren, 1991).

Ozone concentrations can be enhanced in the plume of an industrial plant or in an urban plume at a typical distance of 100 km from the source. Local effects affect the spatial variation of ozone significantly with a coupling between the urban or local and the regional scales. It was found that in urban areas in northern Europe, both NO_x and VOC control could reduce the concentrations of ozone (London urban plume study, Derwent and Hov, 1979). Not much information exists for southern European cities. In a model study of the role of industrial chlorine (Cl_2) emissions within an urban and industrialised airshed, Hov (1985) showed that Cl atom attack on nonmethane hydrocarbons from vehicle exhaust and other sources, can speed up appreciably the generation of secondary products like O_3 and PAN.

a. Topography and urban heat sources influence dispersion and transport

Several studies have shown that the distribution of ozone may be altered both in hilly terrain and in coastal areas by valley and mountain flows and sea and land breezes (Lalas *et al.*, 1983). In large urban areas, the wind can be modified by aerodynamic effects such as changes in surface roughness or by thermal effects caused by land use differences and heat and water vapour rejection (Oke, 1987).

In TRACT, studies have been carried out of the boundary layer growth (Kalthoff *et al.*, 1994) and the vertical distribution of chemical species in complex terrain and under the influence of convective motion (Vögtlin *et al.*, 1994). A data set on transport and diffusion over complex terrain will soon be available. A second data set on tracer transport has been obtained by the TRANSALP section of TRACT (Ambrosetti *et al.* 1991, 1994a, 1994b). The results should help in understanding how the ozone formed in the Po valley and advected towards the Alpine valleys (Sandroni *et al.*, 1994) interacts with the alpine barrier and perhaps passes over to the north. More field studies are needed to test the theoretical understanding, but the information collected can improve existing models and help to establish new model concepts.

Nested grid simulations can be done with the EUMAC Zooming Model either in its stand-alone version or together with the EURAD model (see section 3.3.1 and Chapter 4) as well with other models.

Some conclusions may be drawn about structure of the urban boundary layer from individual contributions to TOR and TRACT, but a number of questions remain unanswered, such as how pollutant concentration may influence the structure of the atmospheric boundary layer through their impact on radiative transfer, clouds and precipitation (Moussiopoulos and Sahm, 1994).

b. Urban anthropogenic and biogenic emissions of air pollutants

In the photochemistry on local and urban scales, the emissions of air pollutants with quite short atmospheric lifetimes like NO_x and some of the non-methane hydrocarbons (NMHC) make the most contribution to photo-oxidant formation.

Emissions are mostly obtained as annual averages for regions or countries, and much less is known about the variability in space and time on the local or urban scale. The quantification of the emissions related to the evaporation from industrial and non-industrial solvent use, exhaust emissions and evaporation from road traffic, biogenic emissions and emissions from power plants, has been seen as particularly uncertain.

The chemical processes which may be of particular importance in the urban atmospheric environment, such as the formation of NO_2, HONO and organic and inorganic nitrate at night, were not addressed in EUROTRAC, but models developed in EUMAC may be used to describe such processes. However, relevant observational data should be collected to confirm the reliability of the model concepts.

3.2.5   Regional or continental photo-oxidants

In the boundary layer, ozone concentrations far in excess of the concentrations aloft, can be formed over rural areas from anthropogenic emissions of nitrogen oxides and volatile organic compounds under the influence of sunlight in high pressure situations. Observational evidence for long-range transport of elevated concentrations of ozone in the atmospheric boundary layer in Europe was first reported in the United Kingdom in the 1970s (Atkins *et al.*, 1972; Cox *et al.*, 1975). Later it was shown theoretically that ozone or its precursors could be transported over long distances in the atmospheric boundary layer (Hov *et al.*, 1978; Eliassen *et al.*, 1982).

The highest summertime concentrations of ozone over Europe are found in the southeast of the continent, with lower concentrations towards the northwest. There is a wintertime minimum in the concentrations while the maximum is usually in early summer. This distribution of photo-oxidants in Europe was analysed in the OXIDATE project supported by OECD. The accuracy of the measurements was stated as better than 20 % of the true value, but this was not established with intercomparisons (Grennfelt *et al.*, 1987, 1988).

Analysed or prognostic information about winds, precipitation, stability and clouds is generally available from the observational network run by the weather services, and from numerical weather prediction models. The dispersion, transport and deposition of photo-oxidants on the synoptic scale has thus been reasonably well understood but it was appreciated that emission data, meteorology and surface information should be better resolved in time and space to obtain a realistic analysis of the ozone distribution over Europe. Dry deposition is a major loss term for lower tropospheric ozone, and the mechanism and rate coefficients for the surface removal of ozone have been studied since the 1970s (Garland and Derwent, 1979). Recent deposition measurements (*e.g.* Jaeschke *et al.*, 1994) have improved the parameterisation of the surface resistances for air pollutants and thus made the estimation of pollutant transport distances more reliable.

The modification of the chemical composition of the atmospheric boundary layer by convective activity has only recently been explored; it contributes a large uncertainty to the quantification of the ozone concentrations since convective clouds appear during periods when the highest ozone concentrations are found (Dickerson *et al.*, 1987).

a. The climatology of ozone and other photo-oxidants in Europe, in particular in southern, central and eastern parts

The location of the measurement stations in the TOR project should show the differences between the climatology of ozone and its precursors within Europe compared to adjacent areas (Cvitaš and Kley, 1994). The measurements show that there is excess ozone in the boundary layer over Europe in the summer, and there is a wintertime ozone reduction. This is readily seen when the measurements at the

Arctic site in Ny Ålesund on Spitzbergen at almost 79 °N are compared with measurements taken 2000 km further south at the mainland station Birkenes not far from the Norwegian south coast, and also with measurements at Mace Head on the coast of western Ireland. The annual cycles of the daily hourly ozone maximum for four years (1988-1991) for the Ny Ålesund site and the Birkenes site are shown in Fig. 3.5. At Ny Ålesund, there is very little annual variability with a small minimum in May, at the other two stations there is an overall decline in concentration in summer and an annual minimum in July. The scatter in the daily maximum ozone concentrations in May reflects the occurrence of surface ozone depletion in April–May at many Arctic coastal sites (Barrie *et al.*, 1988 and Solberg *et al.*, 1994a). At Birkenes there is a November minimum of less than 30 ppb, and a May–June maximum of 50 ppb (Hov and Stordal, 1993). The amplitude in the annual variation of ozone is about 10 ppb at Mace Head (Simmonds, 1993). The seasonal variation and mean concentrations are comparable at Mace Head and Ny Ålesund when the cases with boundary layer ozone loss in the Arctic spring are omitted. The ozone concentration at Mace Head has a maximum in April for unpolluted air identified by CFC11 concentrations below the average value, and there is an annual minimum in July.

Further quantification of the ozone deficit over Europe and surplus in summer was provided by Beck and Grennfelt (1994) who found that in summer, based on the 1989 measurements, there is a gradient in the average diurnal maximum ozone concentration with lower values (30–40 ppb) in the north-western part and higher concentrations (60–70 ppb) towards the south-eastern part of the European network. On average there is a deficit of ozone over Europe in winter (October–March) which is 0–5 ppb near the north-west coast and increases to about 10 ppb in the south-eastern part of the network, with deficits up to 20 ppb in central Europe where the concentrations of NO_x near the surface are high in winter. Beck and Grennfelt *et al.* (1994) identified four sites (Mace Head in Ireland, Svanvik and Jergul in northern Norway, and Strath Vaich in the UK), denoted as reference sites, where the diurnal variation in ozone concentration was very small and the summertime and wintertime daily ozone concentration averages were similar (32 and 31 ppb, respectively). These sites provide data on the composition of clean air in the marine atmospheric boundary layer upwind of the European continent. Several studies in the subproject TOR provide information about the ozone level expected for zero anthropogenic NO_x and VOC emissions. When the relationship between organic $C_3–C_8$ nitrates and O_x (= O_3 + NO_2) was studied in the Schauinsland measurements and extrapolated to zero concentration of the organic nitrates, an ozone concentration of 20–30 ppb was found (Flocke *et al.*, 1994). This value appears to represent the sum of the contributions from stratospheric intrusions and photochemical ozone produced from CH_4, CO and NO_x, since $C_3–C_8$ nitrates are not formed from CH_4 and CO.

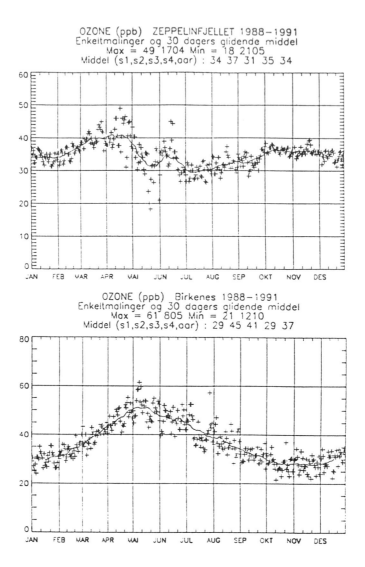

Fig. 3.5: Daily maximum of hourly ozone for 1988–1991 at the Zeppelin Mountain in Ny Ålesund (in ppb), upper part and at Birkenes, lower part. The full line is the 30 day running mean. The average concentrations for January–February–March (JFM), AMJ, JJA, OND and the year are 34, 37, 31, 35 and 34 ppb, respectively, at the Ny Ålesund site at Birkenes the same averages are 29, 45, 41, 29 and 37 ppb, respectively (Hov and Stordal, 1993).

Carbon monoxide is an anthropogenic tracer in air masses where the emissions are much younger than the chemical lifetime of CO (about three months). When O_x and CO are measured simultaneously at a site downwind of man-made emissions, the slope of the linear regression of O_3 versus CO can be used to estimate the amount of ozone formed from the precursors emitted into the air masses together with CO. In winter the slope is often negative indicating a loss of O_x (Sladkovic *et al.*, 1993; Scheel *et al.*, 1994). Penkett *et al.* (1994) measured CO and O_3 on a specific day in December and found a slope of approximately 1:6 in the concentration ratio of O_3 to CO. The O_3 concentration was extrapolated to approximately 15 ppb at zero CO, and this number was interpreted as the ozone present in the surface troposphere due to stratospheric intrusions, a value in good agreement with the available historical data at sea level.

b. Role of NO_x and hydrocarbons to the observed photo-oxidant distributions over Europe

In the TOR subproject, the question "How many ozone molecules are formed in an air mass for each NO_x molecule emitted into it?" has been analysed by Hov (1989) and by Volz-Thomas *et al.* (1993). Measurements at Schauinsland in southern Germany show that in summer, approximately five O_3 molecules are formed per NO_x molecule oxidised (Fig. 3.6), while in winter there is a negative relationship between O_x and the oxidation products of NO_x, supporting the destruction of NO_x at the expense of O_3.

On the basis of observed behaviour of ozone in the rural ozone network over the United Kingdom, Derwent and Davies (1994) calculated that about six ozone molecules were formed for each NO_x molecule oxidised in the summer, while in the winter there is nearly a one to one relationship between O_3 and NO_x loss, thus indicating that the reaction $NO + O_3 \rightarrow NO_2 + O_2$ with further oxidation of NO_2 to NO_z, ($NO_z = NO_y - NO_x$) can account for most of the ozone chemistry in the winter months. In a theoretical calculation of the ozone formation in an air mass in the atmospheric boundary layer passing from Germany to Ireland, Derwent and Davies found that on average 6.2 molecules were produced per NO_x molecule oxidised. It was also shown that the efficiency of ozone production increased with decreasing NO_x levels, in the sub-ppb range; the calculations showed that about 20 ozone molecules were produced per NO_x molecule oxidised. The experimental findings in TOR of higher radical concentrations than predicted by modelling point to a validation problem and should be kept in mind when model results are discussed. A picture of the role of hydrocarbons and nitrogen oxides in photochemical ozone formation in Europe is shown in Fig. 3.7.

From simultaneous measurements of organic nitrates, ozone, NO_x and VOC made at Schauinsland in the plume of Freiburg, it was shown that the fraction of the smaller organic nitrates (C_3 and C_4) was larger than that expected from laboratory data. The oxidation of VOC in these air masses leads, on average, to about four peroxy radicals for each hydrocarbon molecule oxidised by OH, when possible contributions from biogenic VOC are neglected. Consequently, more ozone appears to be produced in a shorter time than is predicted by photochemical models. Preliminary results from laboratory studies seem to confirm the hypothesis

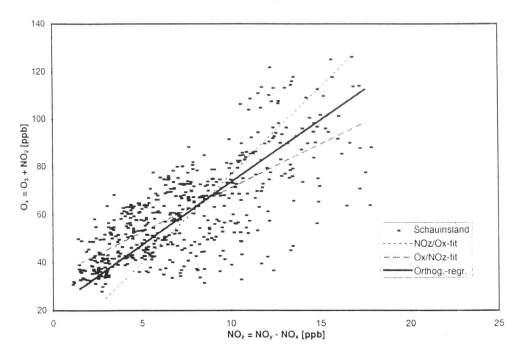

Fig. 3.6: The observed relationship between the oxidant concentration ($O_x = O_3 + NO_2$) and oxidised NO_x ($NO_z = NO_y - NO_x$) at the TOR station, Schauinsland for the wind direction from Freiburg (north-west) in summer (Volz-Thomas *et al.*, 1993). Freiburg is located approximately 10 km away. In summer, approximately 5 O_3 molecules are formed per NO_x molecule oxidised to NO_z, while in winter there is a negative relationship observed between O_x and NO_z which supports the notion that destruction of NO_x occurs at the expense of O_x. The sum of all species derived from NO_x is defined as NO_y, while NO_x is the active form of NO_y which serves to produce ozone through the oxidation of NO by peroxy radicals followed by photolysis of NO_2. The oxidation products, NO_z, are mainly HNO_3, NO_3^-, organic nitrates and N_2O_5. The amount of NO_z is a measure of the fraction of NO_x that is no longer in the active form to produce ozone (based on Voltz-Thomas *et al.*, 1993).

that the large fraction of small organic nitrates is a consequence of the rapid decomposition of the peroxy radicals from the oxidation of larger hydrocarbons.

These findings are in agreement with the measurements of the decay of anthropogenic hydrocarbons and NO_x made between Freiburg and Schauinsland. The measurements were used to derive average OH concentrations during a transport time of about 3 h. There were 5–8 × 10^6 molecules/cm^3 at about mid-day despite the presence of high NO_x concentrations (up to 70 ppb). The sink for OH by reaction with NO_2 could only be balanced if radical amplification was postulated, which would be consistent with the results from studies of organic nitrates (Kramp *et al.*, 1994). There remains the possibility that the additional peroxy and OH radicals come from the decomposition of biogenic VOC, such as terpenes, which would then require smaller amplification factors for the anthropogenic VOC. However, the decomposition of the biogenic VOC would have to produce C_3 and C_4 alkylperoxy radicals and HO_2 at the rate required to explain the findings from the organic nitrates. The large OH concentrations lead to a faster removal of NO_x than is currently assumed in photochemical models, which means that the regime where ozone formation is controlled by the concentration of NO_x is quickly reached. From combined measurements of NO_x, NO_y, H_2O_2 and O_x, it was indeed found that the photochemical formation of ozone at Schauinsland is in most cases limited by the availability of NO_x, even in polluted air from the near-by city of Freiburg (Flocke *et al.*, 1994a,b), which means that an addition of NO_x to the air mass would increase the production of ozone more than a corresponding addition of hydrocarbons.

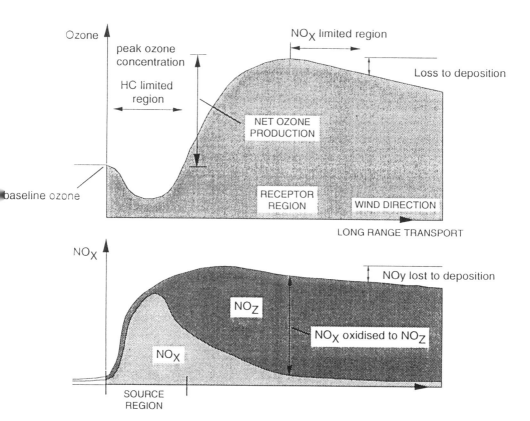

Fig. 3.7:   The relationship between net ozone production and the amount of NO_x oxidised to NO_z (Derwent and Davies, 1994).

It has been noted that there is a local, natural contribution to the concentration of alkenes at some TOR measuring sites. This has been observed, for example, at Schauinsland (Volz-Thomas *et al.*, 1993) and at Birkenes (Solberg *et al.*, 1994b). The measurements at Schauinsland have lead to the conclusion that biogenic emissions of olefins, such as ethene, propene, 1-butene and 1-hexene, may have a considerable impact on the total VOC reactivity in air that is not directly influenced by anthropogenic emissions (Klemp *et al.*, 1993). The contribution of the biogenic hydrocarbons to ozone formation has not yet been quantified.

c. Contribution of free tropospheric ozone and precursors to the average and peak levels of ozone and other photo-oxidants in the atmospheric boundary layer

Today about 31 or 32 ppb represents the atmospheric boundary layer concentration of ozone at the western edge of Europe in westerly air flows. On a seasonal basis, boundary layer processes may add 30–40 % to this value in summer and subtract about 10 % in winter, while in photo-oxidant episodes hourly values of 100 ppb or more can be reached as a result of photochemical formation of ozone within the boundary layer. In episodes, the tropospheric background can be seen as a basis upon which the episodic concentrations of ozone are superimposed, Fig. 3.8 (PORG, 1993). The ozone concentration in the atmospheric boundary layer is determined by exchange of ozone between the boundary layer and the free troposphere, the dry removal at the surface and the *in situ* chemical production and loss. During photochemical episodes in the atmospheric boundary layer, the ozone concentration can exceed the lower free tropospheric concentration so that there is a net flux of ozone upwards out of the boundary layer, but for most of the time the main source of ozone in the boundary layer is a flux downwards from the free troposphere.

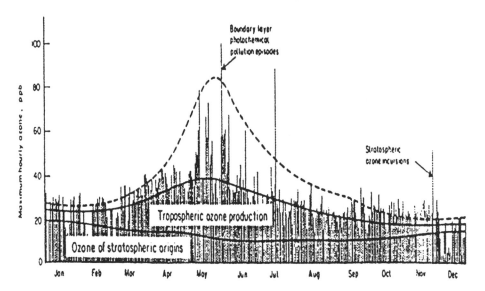

Fig. 3.8: Contribution from the major ozone sources to the daily maximum hourly mean observed at ground level at a rural site in the UK (Derwent and Kay, 1988).

For other photochemical oxidants such as peroxyacetyl nitrate (PAN) and NO_2, as well as the precursors of ozone, the free tropospheric concentration is generally much lower than in the polluted atmospheric boundary layer. Thus the flux of these compounds is upwards from the boundary layer rather than downwards from the free troposphere.

The amount of ozone formed downwind of the source region is controlled by the following factors: the amount of NO_x picked up over the source region, because this determines the upper limit to the amount of NO_x which can be oxidised; the atmospheric chemistry, because this fixes the number of ozone molecules produced per NO_x molecule oxidised at 4–6 for boundary layer average NO_2 concentrations about 1 ppb; hydrocarbons, because their presence drives up the ratio of the number of ozone molecules produced per NO_x oxidised from about 3 for CO to above 6 in air masses which contain significant amounts of methane and man-made precursor emissions; ozone dry deposition, because this process determines the transport distance over which ozone can be transported; sunlight because this determines the spatial scales for ozone production and NO_x oxidation. The greater these production rates, for given precursor emissions, the less ozone and oxidised NO_x are exposed to removal processes, and the higher their concentrations downwind of the source region (Derwent and Davies, 1994).

A model of the exchange processes between the atmospheric boundary layer and the free troposphere by convection has been developed within TOR (Beck *et al.*, 1994). The calculations are based on ozone sonde measurements at Uccle in Belgium. The calculated flux is highly episodic, characteristic for convective events, and has maxima in the summer and low values in the winter. If the net upward convective flux for Uccle is assumed to be representative for continental north-western Europe, there is an upward flux of 1.0 Gmol/d, which is about 5 % of the downward flux of ozone from the stratosphere to the troposphere over the northern hemisphere. Convection is an important process for vertical transport over the continent in summer, but vertical transport is also significant in large scale rising or sinking cells, in frontal zones and over mountainous regions. The vertical transport of precursors, notably NO_x, out of the boundary layer is another process which needs to be included since this intensifies the *in situ* tropospheric ozone chemistry as shown by Chin *et al.* (1994) for North America.

On sunny summer days in central Europe the chemistry is calculated to contribute the dominant term in the change in concentration of ozone in the atmospheric boundary layer with a net generation of about 10 ppb/h on some days (Hass *et al.*, 1993). At locations such as southern Scandinavia, which are far downwind of the main source regions, photochemical episodes with high ozone concentrations can often be observed, but the local chemical generation of ozone is usually low because the precursors are depleted. The local removal terms are mainly horizontal advection, convection and dry deposition. The convective loss is a dominant local loss term when it occurs (up to 7 ppb/h). These calculations were carried out with a coupled 3-D numerical weather prediction and chemical transport model for Europe and the North Atlantic (Flatøy *et al.*, 1995).

d. Dependence of the source-receptor relationships on topography and land use

Topography and surface properties are of particular significance for the source-receptor relationships in mountainous regions, in regions where the air flow is channelled in valley systems and in regions where sea-breeze or larger scale thermally driven circulation systems develop. In the Mediterranean the circulation during the summer is frequently thermally driven during the day with a sinking motion over the sea, and convection over Spain, Italy, Greece, the Croatian coast and the major islands. The circulation is strong and can reach far inland. During the night, the flow is weaker and in the opposite direction. In this way large circulation systems develop where pollutants can be confined for more than a day before exchange with the free troposphere takes place through convective activity. The worst photochemical oxidant episodes in the Mediterranean are probably linked to sub-synoptic scale land-sea breeze circulation systems, affecting in particular the regions with major cities like Barcelona, Marseilles, Rome and Athens (Moussiopoulos, 1994b).

In stagnant air or sea breeze conditions more than 175 ppb of ozone was measured over central Athens in September 1987 (Cerutti *et al.*, 1989, see also Moussiopoulos, 1994b). Bonasoni *et al.* (1991) reported very high ozone concentrations in sea-breeze situations near Ravenna in the Po Valley close to the Adriatic Sea, with particularly high concentrations over the sea (more than 100 ppb of ozone in the afternoon) and with summer average values during the afternoon of more than 70 ppb of ozone. In the Mediterranean large scale circulation cells are established particularly in summer, and coastal emissions can be trapped for several days in the land-sea breeze. During the day thermal lows are formed over the central part of the Iberian Peninsula with a strong upward motion, and it has been suggested that this may be an important mechanism for bringing ozone and its precursors into the free troposphere (Millán 1992, 1994).

In the field measurements carried out in subproject TRACT in the Rhine Valley and on the slopes of the Black Forest (Güsten *et al.*, 1993), large differences were found in the concentrations of ozone as well as in its vertical eddy fluxes. These were attributed to the complex topography and meteorological differences where land-use plays a role. Airborne measurements of ozone and NO_x in southern Switzerland and northern Italy have shown how channelling of polluted air can give rise to concentration differences in ozone of 40 ppb or more in the atmospheric boundary layer over a horizontal distance of less than 20 km (Prevot *et al.*, 1994). Thus source-receptor relationships for photo-oxidants can be strongly influenced by topography in regions with channelling effects.

3.2.6   Photo-oxidants on a global scale

The global aspects of the man-made influences on photo-oxidants have been assessed by the international scientific community under the organisation of UNEP and WMO. The assessments originally addressed the question of changes in stratospheric ozone, but tropospheric ozone and oxidising capacity are intrinsically related to stratospheric ozone depletion and climate change and have increased in importance in these documents (WMO, 1990a; 1990b; 1992; 1994).

The sources of ozone in the free troposphere are transport from the stratosphere and from the atmospheric boundary layer both over densely populated regions and over regions of biomass burning. Photochemical reactions are a sink for ozone, unless NO_x concentrations exceed a critical level, which depends on altitude, water vapour and the ozone concentration itself. Another sink for free tropospheric ozone is exchange with the atmospheric boundary layer in regions where NO_x levels are low due to an enhanced photochemical destruction rate because of the higher water vapour concentrations. Tropospheric ozone plays a dual role in climate: it is a greenhouse gas itself, and as a precursor of OH radicals it influences the removal of other greenhouse gases. For this reason and because the concentration of OH is not linearly related to the ozone concentration, the net effect of changes in tropospheric ozone on climate is not easy to predict (IPCC, 1994).

a. Is there evidence for a significant anthropogenic contribution to free tropospheric ozone concentrations?

The contribution of anthropogenic precursors to the photochemical production of ozone in the summer at northern mid-latitudes has been indicated by satellite data (Fishman et al., 1992), which identify regions of high ozone concentrations that cover large parts of the North Atlantic during summertime. Also, plumes of high ozone were identified in the tropics. There is strong experimental evidence that ozone levels in the southern hemisphere and in the tropics are influenced by emissions from biomass burning (WMO, 1994). However the magnitude and the secular trend of man-made burning with respect to southern hemispheric ozone levels are uncertain. No trend is seen in the surface ozone records obtained over the last two decades at Cape Point, South Africa (Scheel et al., 1990) and American Samoa in the Pacific Ocean (Oltmans and Levy, 1994). Andreae (1994) estimated that the release of trace gases from biomass burning has increased by a factor of two or three since 1850. Semi-quantitative measurements made during the last century, although not considered sufficiently reliable for an independent quantitative assessment (Kley et al., 1988), support a secular increase in tropospheric ozone concentrations in the southern hemisphere (Sandroni et al., 1992).

Transport of anthropogenic pollutants to upper levels near the tropopause may occur through strong convection at most latitudes. For example Ehhalt et al.

(1992) estimated that about 30 % of the NO_x found in the upper troposphere may result from upward convective transport at some latitudes.

An appreciable fraction of NO_x may also be also be generated by air traffic in the flight corridors at the level of the tropopause (Schumann, 1994). The impact of air traffic on ozone formation in the upper troposphere and lower stratosphere has been investigated with EUROTRAC models. The ozone increase attributable to air traffic within the corridors may be as much as 10 % of the background ozone (Zimmermann, 1994, Strand and Hov, 1995), the magnitude depending on the meteorological conditions at the time of emission (Petry *et al.*, 1994, Ebel and Petry, 1994, Flatøy and Hov, 1995).

b. What is the contribution of exchange with the stratosphere to ozone concentrations in the free troposphere?

Downward transport from the stratosphere is one of the main sources of ozone in the free troposphere. The most active regions of stratosphere-troposphere exchange are near jet streams in cyclonic regions of the upper troposphere and in cut-off lows. It was a EUROTRAC contribution which succeeded for the first time in applying a mesoscale simulation model to calculate the ozone fluxes from the stratosphere to the troposphere within tropopause folds and cut-off lows (Ebel *et al.*, 1991). Detailed examination of the processes occurring led to the conclusion that such events may strongly intensify cross-tropopause ozone fluxes. Another important finding is that downward air-mass fluxes are partly compensated for by upward fluxes (Ebel *et al.*, 1995). A climatology of foldings and cut-off lows has been provided for global cross-tropopause air mass and ozone-flux estimates in the framework of the EUMAC (Ebel *et al.*, 1993, 1995). The climatology is in agreement with findings by other authors that cross-tropopause transport is about twice as effective in the northern hemisphere as in the southern. A particular result is the identification of pronounced longitudinal differences at northern latitudes.

The contribution of tropopause folds to the exchange has been confirmed by recent TOR work (Beekman *et al.*, 1994) and the role of extra-tropical cyclones was emphasised in a GLOMAC study (Alaart *et al.*, 1994). The decrease in tropospheric ozone at the South Pole (Oltmans and Levy, 1994) is likely to be a consequence of the large ozone depletion in the stratosphere over Antarctica. However estimates of fluxes across the tropopause remain uncertain. Murphy and Fahey (1994) used measurements of the ratio between ozone and NO_y and an annual destruction rate of N_2O in the stratosphere of 8–17 Tg(N)/a to infer a transport of 0.28-0.6 Tg(N)/a of NO_y and 240-820 Tg/a of ozone into the troposphere. This is slightly less than the earlier estimates made from observations of tropopause folding events (Danielsen and Mohnen, 1977), but is comparable to the fluxes derived from global circulation models (*e.g.* Gidel and Shapiro, 1980; Levy *et al.*, 1985). There have been no studies of trends in stratosphere-troposphere exchange, so the contribution of the stratospheric source to the observed trend in tropospheric ozone remains an open question (WMO, 1994).

c. What controls the chemical balance of ozone in the free troposphere of the chemical balance?

In remote areas of the troposphere, CO and methane provide the fuel for ozone production (Seiler and Fishman, 1981), since these compounds are the main reactants that convert OH radicals into HO_2 and RO_2. The more reactive hydrocarbons are of lesser importance because of their much lower abundance. For example, the long-term measurements made at the Izaña, Tenerife TOR station (Schmitt et al., 1994) show that CH_4 makes up most of the total reactivity of VOC. The critical component in the net ozone balance is the concentration of NO, the essential catalyst in ozone formation (Crutzen, 1979).

In the remote atmosphere, not only is the production rate of O_3 limited by the availability of NO_x, the concentration of NO_x can be so small that photochemical reactions consume ozone (Liu et al., 1983). This was confirmed by recent measurements in remote regions of both hemispheres (Ayers et al., 1992; Ridley et al., 1992; Smit and Kley, 1993). However, even very small concentrations of NO_x contribute to O_3 production, compensate the loss rate, and increase the lifetime of O_3. Since the chemical production of ozone is sensitive to the NO_x abundance while the loss in the remote atmosphere is not, the concentrations of ozone in the free troposphere depend upon the concentrations of NO_x (WMO, 1994).

d. Which factors control the budget of NO_x?

The main sources of NO_x in the free troposphere are lightning, transport from the stratosphere, aircraft emissions and transport from the atmospheric boundary layer. In WMO (1994), it was estimated that aircraft emissions can raise the upper tropospheric ozone levels by 3–10 %. The large uncertainty reflects the uncertainties in estimating the other NO_x sources in the upper troposphere. It was noted that more research is needed of the source strength and spatial distribution of NO_x from lightning, and the distribution and chemical conversion of NO_x during convection.

Other problems in assessing the budgets of ozone and NO_x in the free troposphere are associated with the processes that determine the lifetime of NO_x in the different parts of the atmosphere and, hence, control the efficiency of NO_x in catalysing ozone formation. NO_x acts as a catalyst in ozone formation until it is permanently removed by deposition, either in the form of NO_2 or after chemical transformation into other NO_y species, particularly HNO_3 or nitrates. The lifetime of NO_x varies from a few hours in the boundary layer to several days in the upper troposphere. During the day, NO_x is converted to nitric acid and peroxyacetyl nitrate (PAN) and its homologues. In the planetary boundary layer, HNO_3 provides an effective sink for NO_x, but the thermally unstable compound PAN provides only a temporary reservoir for NO_2. The lifetime of PAN becomes long enough at the colder temperatures of the middle and upper troposphere allowing it to be transported over long distances, and to carry NO_x into remote regions. NO_x is also removed by the formation of alkyl nitrates ($RONO_2$). As with PAN, alkyl nitrates can provide a source of NO_x in more remote regions through photolysis or through reaction with OH following transport (Atlas, 1988; Flocke, 1992). While HNO_3 is removed relatively quickly by dry and wet deposition in the atmospheric boundary layer, it has a much longer lifetime in the free troposphere. There, a photochemical steady state can be established between NO_x and HNO_3 so that, from a chemical point of view, both persist indefinitely. Losses occur by exchange with the atmospheric boundary layer and by heterogeneous processes.

Another important loss process for NO_x is the oxidation of NO_2 by ozone itself. The NO_3 radical then formed is extremely sensitive to photolysis during the day, but at night it can reach concentrations of several hundred ppt (Platt et al., 1981; Wayne et al., 1991). A thermal equilibrium is established between NO_3, NO_2 and N_2O_5 and therefore heterogeneous loss of N_2O_5 to form nitrate aerosol is a sink for NO_x which is thought to be significant on a global scale (Dentener and Crutzen, 1993).

The few available model studies (Liu et al., 1987; Lin et al., 1988; Hov, 1989; Strand and Hov, 1994, 1995) suggest that the catalytic efficiency of NO_x in making ozone could increase from values around 10 molecules of ozone made for each molecule of NO_x emitted at high NO_x concentrations (see 3.2.5), to values of up to 100 in the remote atmosphere (Liu et al., 1987). The models also suggest that the efficiency could be larger in winter than in summer because of the longer winter lifetime of NO_x. However, while the model results are in reasonable agreement with atmospheric measurements (see 3.2.5) the high efficiency of NO_x as a catalyst in ozone formation in the free troposphere has yet to be confirmed by experimental data.

e. What is the influence of export of ozone and precursors from North America to Europe?

Ozone and its precursors have longer lifetimes in the free troposphere than in the atmospheric boundary layer, and so the vertical redistribution of ozone and its precursors between the boundary layer and higher altitudes has a strong influence on the ozone distribution in the troposphere. Upward flow in convective systems serves to bring ozone and its precursors from the continental atmospheric boundary layer into the free troposphere, where the lifetimes are long enough to allow transport over large distances. The upward flux is, however, balanced by downward mesoscale flow, where ozone and odd-nitrogen species are brought from the free troposphere into the planetary boundary layer where the lifetimes are shorter (Lelieveld and Crutzen, 1994). The exchange between the atmospheric boundary layer and free troposphere is an episodic process and is therefore difficult to include explicitly in global models. The importance of convection for the understanding of free tropospheric ozone has been demonstrated in TOR from the analysis of vertical soundings with a 3-D model (Flatøy et al., 1995).

The impact of North American emissions on the global ozone budget was estimated by Parrish *et al.* (1993), who observed a strong correlation between ozone and CO with a consistent slope $O_3/CO = 0.3$ at several island sites in eastern Canada. Jacob *et al.* (1993) used a 3-D model and the same measurements to estimate that pollution from North America contributes 30 Tg of O_3 to the northern hemisphere in summer, of which 15 Tg is due to direct export of ozone and 15 Tg is due to export of NO_x which leads to O_3 production in the free troposphere. This anthropogenic source of O_3 is about one third of the estimated cross-tropopause transport of O_3 in the northern hemisphere in summer. Since North America accounts for about 30 % of man-made NO_x emissions in the northern hemisphere, it can be concluded that in the northern hemisphere anthropogenic sources contribute an amount comparable to the flux from the stratosphere to tropospheric ozone.

Evidence for the influence of export from North America on ozone levels in the free troposphere over Europe and the North Atlantic is found in the vertical ozone soundings and in the measurements made at high altitude observatories. Fig. 3.9 compares the seasonal cycles of ozone obtained at Mauna Loa in the Pacific with those at the TOR stations Izaña, Tenerife, in the southern North Atlantic, and the Zugspitze in southern Germany. During winter, the ozone concentrations are about 40 ppb and are similar at all three stations. All the stations show a similar increase at the beginning of the year but peak at different times. At Mauna Loa, which is furthest away from populated areas, a maximum of 55 ppb is reached in April. At Izaña and the Zugspitze ozone concentrations continue to rise during the spring and remain at the high level until June and August, respectively.

The vertical soundings from Jülich also show that lower concentrations of ozone are present in the free troposphere in winter (about 50 ppb) than in spring and summer (50–70 ppb; Smit *et al.*, 1993). It is likely that both Jülich and the Zugspitze are already influenced by European emissions. At the Zugspitze, this conclusion is supported by a positive correlation between ozone and CO in summer and a negative correlation in winter (see 3.2.5). However, at Izaña high ozone concentrations persist until August in air masses arriving from the northern part of the North Atlantic (Schmitt *et al.*, 1993). Since Izaña is located on the western boundary of Europe and because most of the air masses arrive from north westerly directions, the higher ozone concentrations in summer are probably the result of export of ozone from the North American continent into the free troposphere over the Atlantic and Europe.

Chemical processes are slow in the upper free troposphere and fast in the marine boundary layer, while the transport is fast in the upper free troposphere and slow in the atmospheric boundary layer. It is not possible to give a clear answer without a detailed model calculation, but the lower ozone concentrations in winter suggest that the catalytic efficiency of NO_x for ozone formation in the free troposphere is lower during the winter.

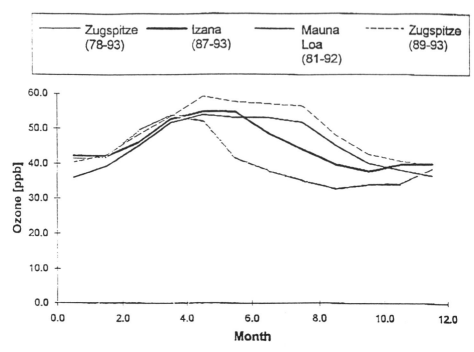

Fig. 3.9:   Average seasonal cycle of ozone observed at the TOR station, Izaña and
compared with those from the Zugspitze on the European continent and Mauna Loa,
Hawaii (Schmitt 1994; Sladkovic *et al.*, 1993; Oltmans and Levy, 1994). All three stations
show an increase in early spring. Ozone in the Pacific at Mauna Loa decreases during
summer, however, while relatively high concentrations persist until August at the
Zugspitze. The concentrations at Izaña fall in between, which is a result of the combination
of low concentrations that are advected from the southern parts of the North Atlantic or the
Saharan Desert, and high concentrations that are advected from the northern part of the
hemisphere (Schmitt *et al.*, 1994 (TOR)). The highest concentrations of up to 80 ppb occur
frequently together with increases in methane and other trace gases that are indicative of
anthropogenic pollution, such as CO, hydrocarbons, and PAN. This is most evident in
spring, when temperatures are cold enough so that PAN can survive long range transport to
Tenerife. It is unclear at present how much the ozone levels at Izaña are influenced by
stratospheric intrusions. This is subject of ongoing research in TOR and in the EU project
OCTA.

Two possible reasons are the destruction of NO_x and ozone through night-time chemistry and cloud processes, and a faster exchange between the free troposphere and the atmospheric boundary layer than was assumed in the earlier model calculations.

The export of ozone and precursors from Europe occurs mainly at the eastern boundary where measurements are not available. It should be of similar magnitude as that from North America. The measurements at Mace Head have been analysed for export of ozone from Europe (Derwent *et al.*, 1994) but they are not really representative for the overall export from Europe, since only a small fraction of ozone and its precursors from European sources leaves Europe in westerly directions. Also Mace Head is located at sea level and cannot provide information on transport in the free troposphere.

f. Do clouds influence significantly the balance of NO_x and ozone?

In clouds there is a separation of HO_2 and NO because HO_2 is very soluble and enters the aqueous phase while NO is insoluble. There is therefore little reaction between HO_2 and NO in clouds. On the other hand, HO_2 in the aqueous phase leads to ozone destruction. Cloud effects have been modelled in GLOMAC, and it appears that the chemical production rate of ozone in the lower troposphere, where most clouds are found, is reduced by 30–40 %, while O_3 destruction reactions are enhanced by up to a factor of two (Lelieveld and Crutzen, 1990). However, the resulting reduction in ozone concentration in the lower troposphere also lowers the O_3 dry deposition flux. In consequence, the global amount of O_3 in the troposphere may be reduced by 10–30 % due to cloud chemistry, when compared with a cloud-free atmosphere (Jonson and Isaksen, 1993; Dentener *et al.*, 1993). Another important property of clouds is to enhance the heterogeneous conversion of NO_2 to nitrate via NO_3 and N_2O_5, as shown for example by Dentener *et al.* (1993).

g. What is the influence of an increase in UV-B radiation due to stratospheric ozone depletion on tropospheric ozone concentrations?

The depletion of stratospheric ozone during the last decade has led to increased ultraviolet radiation of wavelengths 290–320 nm penetrating into the troposphere. An increase in the UV radiation probably gives rise to a reduction in tropospheric ozone in regions of low NO_x where there is a net photochemical loss already; for example over large areas of the southern hemisphere and the remote oceanic regions of the northern hemisphere. On the other hand, an increase in UV radiation will probably lead to an increase in photochemical ozone formation in the NO_x rich continental regions and possibly also in large-scale plumes downwind of these regions. The overall effect of an increase in UV-B radiation on ozone concentrations in air masses that are transported into Europe is not clear.

h. How do changes in ozone and its precursors influence the removal rates of other trace gases that are relevant to climate?

The concentrations of many trace gases, that contribute to the greenhouse effect of the atmosphere (for example CH_4 and HCFCs) or are involved in the atmospheric ozone budget (*i.e.* CH_4, HCFCs and CH_3Br in the stratosphere and CO, CH_4, NMHC and NO_x in the troposphere), are modified through reactions with OH radicals. The concentration of OH is strongly linked to the UV flux below 320 nm (UV-B) and to the concentrations of water vapour and ozone (see 3.2.3). In addition, OH is affected by other trace gases. For this reason, rising levels of CH_4, CO and NO_x may lead to changes in the oxidising capacity of the troposphere (Thompson and Cicerone, 1986; Isaksen and Hov, 1987), which in turn would influence the concentrations of gases relevant to global warming and stratospheric ozone depletion. An increase in UV flux, H_2O and O_3 would immediately lead to an increase in OH, as shown by the good correlation between OH concentrations and the photolysis frequency of ozone (Platt *et al.*, 1988). However the net effect of enhanced UV radiation and H_2O concentrations on OH also depends on the development in the ozone concentration, and therefore on the NO_x concentration, and the transport of ozone from other regions.

Direct measurements of OH concentrations in the troposphere are sparse. Average numbers for global OH concentrations have been derived from the concentrations of tracers, that are removed from the atmosphere by OH. These are CH_3CCl_3 with an atmospheric turnover time of about 6 years and ^{14}CO with a turnover time of a few months. The first estimates from both tracers gave a weighted average OH concentration of 6 to 7×10^5 molecules/cm^3 (Singh, 1977; Volz *et al.*, 1981). New measurements of ^{14}CO, made in the southern hemisphere and at the Schauinsland TOR station, confirmed the earlier estimates of a tropospheric OH concentration of about 7×10^5 molecules/cm^3. The new ^{14}CO measurements suggest a faster removal of ^{14}CO in the southern hemisphere, in contradiction to current model estimates, and much weaker latitudinal gradients than those derived by Volz *et al.* (1981). The new data indicate either a missing sink for atmospheric CO, which is most effective at high latitudes, or a faster meridional exchange, in particular between polar regions and mid-latitudes.

i. What is the overall effect of anthropogenic emissions on climate?

Tropospheric ozone is a greenhouse gas itself and influences, as a precursor of OH radicals, the removal of other greenhouse gases. In the pre-industrial atmosphere, ozone levels were probably 10–20 ppb in the atmospheric boundary layer and somewhat higher in the upper troposphere (see section 3.2.1.), although there is not enough information to give a fully global assessment. The corresponding (unknown) "natural" OH concentration served to keep CH_4 in a steady state with its sources at a concentration of about 600 ppb. Today, CH_4 concentrations are about 1.7 ppm and have been increasing since the beginning of the industrial revolution, apart from the last decade (IPCC, 1994). It is not clear if this increase was simply due to increases in CH_4 sources or if part of the increase was the consequence of a simultaneous decrease in OH concentrations.

In addition to CH_4, a number of other trace gases which have large global warming potentials (GWP) and ozone depletion potentials (ODPs) are released into the atmosphere. Most of these compounds have relatively long atmospheric residence times, a year or more, and are distributed over the entire troposphere or at least one hemisphere. Higher OH concentrations would benefit the removal of these compounds. However, the available data and model calculations show that our current understanding of the global budget of OH is insufficient to make confident estimates of its trend as a consequence of changes in anthropogenic emissions. Changes in ozone are not linearly related to changes in OH. The overall effect of man-induced changes in ozone on climate depends further on the GWPs of ozone relative to the GWPs of the trace gases that are removed by OH. For these reasons, the net effect of changes in tropospheric ozone on climate is not easy to predict (IPCC, 1994). It is clear, however, that the tropospheric concentrations of trace gases like CH_4, CO, HCFCs *etc.* would increase to much higher levels without the presence of ozone in the troposphere.

3.3   Applications and implications for environmental policy

The application of scientific results to environmental policy requires a synthesis of information of how emissions of precursors are transformed, dispersed and removed. Such a synthesis of information is done in simulation models, or in field experiments. In EUROTRAC both these approaches have been followed.

3.3.1   Local and urban photo-oxidants

The characteristic times for chemical transformation and the dispersion of photo-oxidants cover a wide range, and it has been recognised for some time that urban and local scale problems can only be treated successfully with the aid of numerical models, in which either a large enough domain is considered, or accurate boundary conditions are established. The former requires large hardware resources; for the latter, models with nesting capabilities are required, and these have only recently become available. In policy for the local or urban scale, the ozone isopleth method was used in the 1970s to estimate the NO_x and VOC emission control required to attain a certain upper limit for ozone (applied for London by Hov and Derwent, 1981).

A promising approach for addressing environmental policy questions is to apply a model system where the best scientific knowledge is incorporated. Within EUROTRAC, such model systems were developed and applied in EUMAC. The main task of EUMAC is to carry out long-range chemical transport modelling and that is done with the EURAD model system, which also can describe phenomena with characteristic length scales of the order of 50 km (Jakobs, 1994). The EURAD model has been used to investigate the influence of urban plumes on the formation of photo-oxidants over rural areas downwind. One example was the application to describe the transport of polluted air from the Rhine valley and the city of Freiburg to Schauinsland and the TOR station there (Geiß et al., 1994) where for example. the effect of variation of the boundary layer height on transport properties and concentrations of the chemical species can be evaluated.

Two smaller-scale model systems have been employed in EUMAC for local and urban photo-oxidant calculations with the option of using the EURAD data as input: KAMM/DRAIS (Nester et al., 1993) and the EUMAC Zooming Model (EZM). The EZM can be used either in conjunction with the EURAD model or in a stand-alone mode driven with measured data (Moussiopoulos, 1994a). The following account gives some examples of the application of the EZM in policy-related situations. Also listed are some unresolved questions that could be addressed by the EZM or by models such as MAR, which is a coupling of the Colorado State University meso-meteorological model and the American urban airshed model. MAR has been adapted for special applications in coastal zones (Borrego et al., 1994a,b).

Several of the applications of the EUMAC Zooming Model are related to the solution of environmental policy problems. Examples are the cases of Athens, Barcelona and the Upper Rhine Valley.

a. Athens

Athens is a good example of a large city which suffers from extremely high air pollution levels. The Athens' case is also challenging from the scientific point of view because it combines elevated anthropogenic emission levels and complex topographical and meteorological conditions. The Athens Basin is surrounded by mountains on three sides and is open to the sea to the south-west (Fig. 3.10). This often results in sea breeze circulations in summertime anticyclonic situations, when most smog episodes in Athens take place. However, the highest air pollutant concentrations in Athens occur when a weak synoptic forcing acts against the development of a sea breeze and air mass stagnation is established over the city. Nested grid simulations were performed with the EZM to provide an adequate description of the mesoscale wind flow in the Greater Athens area. A coarse grid calculation was used to formulate boundary conditions for the fine grid simulations (Fig. 3.10). Fig. 3.11 shows EZM results for the near-ground wind field and the near-ground concentration patterns of CO and O_x (= NO_2 + O_3) in the Greater Athens area at 18 LST on 25 May 1990. As the synoptic conditions that day prevented the development of the sea breeze, the ventilation over Athens was weak and the air transport between the Athens Basin and the surrounding areas was severely restricted.

Comparisons of model results and observations show that the EZM reproduced the observed situation well. To optimise an air pollution abatement strategy for Athens, the results of simulations with the EZM for various emission scenarios were considered. In particular, it was shown that road traffic emissions could be reduced effectively by introducing

(1) subsidies for new clean passenger cars,

(2) subsidies for retrofitting in-use cars,

(3) the improvement of inspection and maintenance, and

(4) traffic restrictions for polluting vehicles.

It was predicted that this would lead to a substantial decrease of peak hour NO_2 values in the city centre and also to a noticeable decrease of ozone in the periphery of Athens (Fig. 3.12), in spite the expected increase in the vehicle fleet by 40 % in the period between 1990 and 2000. Although only the first measure was introduced, the predictions seem to be confirmed by monitoring measurements in Athens (Moussiopoulos, 1994b).

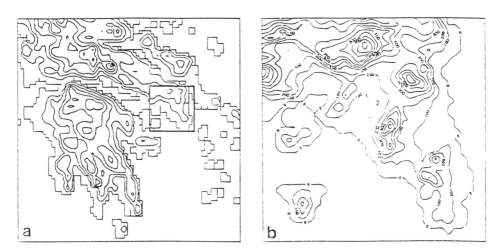

Fig. 3.10:   Coarse grid (a) and fine grid (b) for the calculation of wind flow and the dispersion of chemically reacting pollutants in the Greater Athens Area. The frame in (a) corresponds to the position of the fine grid.

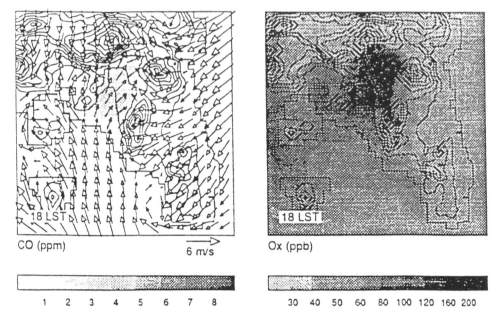

Fig. 3.11:   Results of the EZM for the surface level pattern of the horizontal wind velocity and the concentrations of CO and O_x (= NO_2 + O_3) in the Greater Athens Area at 18:00 LST on May 25, 1990.

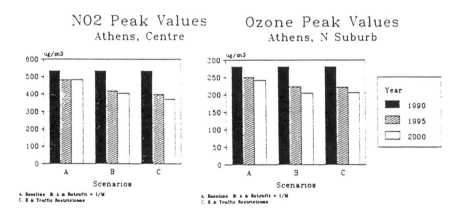

Fig. 3.12:   Peak hourly values of NO_2 in the city centre and of ozone in a northern suburb of Athens predicted for 1995 and 2000 for three emission scenarios compared to the 1990 value. A: intervention 1; B: interventions 1–3; C: all interventions.

b. Barcelona

The results of a series of simulations with the EZM were used to support the decisions taken with regard to traffic regulations in Barcelona during the Olympic Games of 1992 (Baldasano *et al.*, 1993). The EZM was used to investigate the air quality on a summer day in the Greater Barcelona area. The model results for the wind field largely corresponded to the expected diurnal variation, with a land breeze combined with katabatic air motion at night and a deeply penetrating sea breeze during the day. The dispersion simulation shows that, in the course of the night, large amounts of polluted air are transported out to sea, the area there serving as a reservoir in the hours before sunrise. Driven by the sea breeze, this air mass is then transported back to the urban area in daytime forming ozone, while new emissions of primary pollutants strengthen the ozone formation. In this way considerable amounts of photo-oxidants may be transported far inland. The EZM results were in good agreement with the observed diurnal variation of pollutant concentrations during smog episodes in Barcelona.

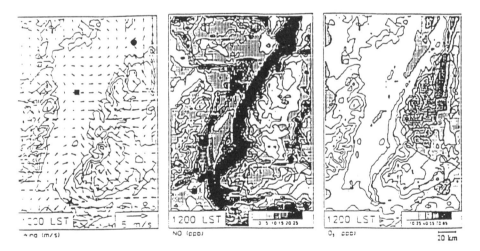

Fig. 3.13:   EZM results for the surface level wind field and the surface level NO and ozone concentrations in the Upper Rhine Valley on September 16, 1992 at 12:00 LST. The location of the cities of Karlsruhe, Strasbourg and Basel is indicated by the symbols:

c. Upper Rhine Valley

The Regio-Klima Project (REKLIP) was launched as an international research project involving German, French and Swiss scientists to create a climate atlas for the Upper Rhine Valley and to establish a dense meteorological measuring network to analyse the regional climate characteristics (Fiedler, 1992). The EZM was applied to the Upper Rhine Valley for one of the days during the REKLIP intensive measuring campaign. Hourly emissions for the model domain were derived from the REKLIP emissions inventory. A comparison between computed and measured wind velocities shows that the change of wind direction in the valley from southerly to northerly during the day is well reproduced by the model. The channelling of the flow during the morning leads to a total depletion of ozone in the interior of the valley. In contrast significant amounts of ozone may persist at elevated regions in the Black Forest because of the absence of local NO sources (Fig. 3.13). A higher VOC/NO$_x$ ratio in the afternoon results in a strong ozone production in most parts of the domain. The predicted diurnal variation of ozone agrees fairly well with observations. The effect of changed anthropogenic emissions was analysed in model sensitivity studies. A reduction in the road traffic outside the urban centre may cause an increase of the ozone levels in the interior of the model domain, but a decrease in the usage of organic solvents can lead to a significant reduction in the ozone levels (Schneider, 1994).

d. Potential EUROTRAC contributions to environmental policy solutions for local and urban photo-oxidants

Many further practical problems could be treated with the methodology and knowledge acquired within EUROTRAC. Here are some examples.

The development of a system for the short-term forecast of air pollution episodes and for the prediction of the long-term development of the air quality in specific urban areas for different air pollution abatement strategies.

The formulation of appropriate methods to be followed in environmental impact assessment studies for air pollution. A harmonisation of the procedure followed in Europe is needed. The level of expertise achieved in EUROTRAC now makes it possible to account properly for the characteristics of individual regions in the continent.

Monitoring networks in heavily polluted areas including urban areas could be optimised. For this purpose, available models would be used to identify in which locations meteorological parameters and pollutant concentrations should be measured in order to characterise the pollution levels for the area of interest and avoid strong local influence.

Clarification of how "local" measures, that is those limited to a specific area of about 100×100 km^2 and valid for a few days may contribute to reducing photo-oxidant levels. In this sense, models developed within EUROTRAC could be used to derive appropriate contingency plans for individual European regions.

Priorities regarding VOC control could be identified. For areas where the control of hydrocarbons proves beneficial locally, criteria are needed to decide on the effectiveness of individual measures. Such criteria are site dependent and have to be formulated so as to take into account both the characteristics of the area and possible regional influences.

An integrated system could be developed for the comprehensive assessment of air pollution abatement policies in specific urban areas, including analyses of the socio-economic aspects. Of relevance here is the scientific programme launched by the European motor and oil industries to identify the influence of fuel composition and advanced vehicle/engine technologies on exhaust emissions, with the aim of assisting the European Union to establish emission limits and fuel specification criteria for the year 2000, on a cost-effective basis linked to air quality requirements (Palmer, 1994).

3.3.2   Regional or continental photo-oxidants

A number of models were applied before 1985 to calculate ozone concentration in the atmospheric boundary layer during summer time episodes (PHOXA, EMEP, ADOM/TADAP, NCAR/RADM). The PHOXA and the EMEP models were used to study several episodes in Europe (Pankrath, 1989; Hov *et al.*, 1985). It was found that in episodes in the atmospheric boundary layer, VOC emission reduction was always effective in reducing peak ozone concentrations, and that the same mostly also held for NO_x. It was also found that VOC or combined VOC + NO_x reductions were more effective than NO_x reductions alone. For regions outside the UK, the Netherlands, Belgium and parts of Germany, the emissions of NO_x were found to affect the ozone concentration more than the VOC emissions. Models that could calculate the long-term average ozone concentration were also available or under development. In Isaksen and Hov (1987) an example was given of long-term calculations of ozone changes in the troposphere based on climatological transport information. It was shown that ozone in the free troposphere could increase as the emissions of NO_x, CO and CH_4 increase. These calculations were important for understanding how surface emissions may alter an important boundary condition for surface ozone: the ozone concentration in the lower free troposphere. There was much room for improvement, for example through a description of the exchange between the boundary layer and the free troposphere.

a. Distribution and trends for photo-oxidants in the atmospheric boundary layer in Europe

The scientific assessment in section 3.2.5 leads to the following conclusions which are important in policy making: in summer the average diurnal maximum ozone concentration is 30–40 ppb in the north-western part and 60–70 ppb towards the south-eastern part of Europe. In winter (October to March) on average there is a deficit of ozone over Europe which is 0–5 ppb near the north-west coast, increasing to about 10 ppb to the south east, with deficits up to 20 ppb in central Europe where the concentrations of NO_x in winter are high near the surface. The background ozone concentration upwind of Europe in the marine boundary layer is quite constant over the day, and the summer and wintertime averages are similar (32 and 31 ppb, respectively).

For volatile organic compounds, which are precursors for ozone, there is a statistically significant upward trend in the concentrations of acetylene, propane, butane and also the sum of C_2–C_5 hydrocarbons, and a downward trend in the concentration of alkenes (ethene and propene) at the TOR site with the longest record of hydrocarbon measurements (Birkenes in Norway, since 1987).

An application of the analytical procedures used in TOR took place when the methods for determining VOC in air samples, developed, tested and put in operation at TOR sites, were later adopted by EMEP when it was decided to include VOC measurements at some of their sites. The EMEP sites chosen for VOC measurements were selected in co-operation with the TOR participants in order to obtain a better picture of VOC concentrations over Europe.

b. Man-made contribution to the observed distributions and trends

A finding, important for ozone control policy, was made in TOR. Combined measurements and model calculations indicate that the photochemical production of ozone, even in an urban plume like the one from Freiburg in southern Germany, is, in most cases, limited by the availability of NO_x. This means that the addition of NO_x to the air mass would increase the production of ozone to a larger extent than a corresponding addition of hydrocarbons (Flocke et al., 1994).

The ozone generation in the boundary layer over the United Kingdom was calculated from the observed behaviour of rural ozone across the country. It was shown that about 6 ozone molecules were formed for each NO_x molecule oxidised in the summer, while in the winter there was a nearly one to one relationship between O_3 and NO_x loss, indicating that the reaction $NO + O_3 \rightarrow NO_2 + O_2$ and then the further oxidation of NO_2 to NO_z can account for most of the ozone chemistry in the main winter months (Derwent and Davies, 1994; PORG, 1993). These results are supported by experimental results obtained in TOR.

Model calculations with the Dutch LOTOS model (Builtjes, 1992) and the EMEP model (Simpson 1993, 1995) have shown that there is no linear relationship between the ozone concentration and the VOC and NO_x emissions. There is, however, a nearly linear relationship between the change in the ozone concentration in a receptor grid square in the model domain and the change in the man-made VOC emission in an emitter country over a broad range of values of VOC emission changes.. The ratio of the ozone change to the change in VOC emission varies by almost a factor of five among the 27 European countries (highest ratio for the United Kingdom and lowest for Norway) (Simpson, 1995). The conclusions from the field experiments in the Freiburg urban plume need to be kept in mind when evaluating model calculations. It was found there that NO_x was converted faster than expected from existing models so that an NO_x limitation for ozone formation was established not far downwind from the urban NO_x source.

Accurate emission data are essential for policy-oriented model calculations. EUROTRAC has contributed to the improvement of emission data used by EMEP with significant progress being achieved recently. EMEP has largely concentrated on annual emission data, and applied relatively simple assumptions on variation with time. The better knowledge on the variation of the emissions obtained in GENEMIS has been used in the EUMAC models and is expected to be included in the EMEP models as soon as European-wide procedures become available.

Results from LACTOZ have been applied by EMEP through the Chemical Mechanism Working Group, where the reaction scheme used in the EMEP photo-oxidant model was reviewed and some revisions proposed. In this way recent results on reaction rate coefficients and reaction pathways from LACTOZ and from international photochemistry re-evaluations in general, have rapidly become available in an evaluated form to the policy oriented applications (Wirtz *et al.*, 1994). An intercomparison study is presently underway of the EURAD model (EUMAC), the LOTOS model (TOR), the REM-III model and the EMEP model. The aim is to investigate the influence of increasing model complexity and detail on the performance. Calculations have been carried out for a particular time period during July and August 1990, and comparisons made of the calculated concentrations of O_3, NO, NO_2, PAN, H_2O_2 and HNO_3 both between the models and with experimental measurements at six sites. The influence of model parameters such as dry deposition velocity and of model input such as biogenic VOC emissions, is being investigated. The results will be published early in 1996.

Model calculations with both the LOTOS and the EMEP models show that the effect of the VOC emission reductions depends on the NO_x emission level. NO_x emission reduction always gives rise to reduction in the ozone concentrations outside urban regions. It can be seen from Fig. 3.14 that, regardless of NO_x, concentration, a 50 % VOC emission reduction across the EMEP grid gives a positive ozone reduction in all grid squares, while for a 50 % NO_x emission reduction, there is an ozone reduction in all but five of the 709 grid squares for which calculations were made; in these five the NO_x concentration is particularly high (10 ppb or more of NO_x) (Simpson, 1991). When the NO_x emissions from man-made sources in Europe were fixed at a lower level, the effect of VOC emission reduction on the concentration of ozone was in general reduced by the same factor. The ozone concentrations on a regional scale in Europe have been shown in these model calculations to be controlled both by the emissions of NO_x and of VOC, although the experimental results from Schauinsland, Freiburg (Flocke *et al.*, 1994) tend to emphasise NO_x emissions more.

In southern Europe, local formation of photo-oxidants is often found in connection with local scale atmospheric circulation situations (land-sea breeze), and emissions of VOC (including biogenic VOC) may be of special importance for the peak ozone concentrations there. In general the contribution of the biogenic hydrocarbons to ozone formation in Europe has not yet been quantified.

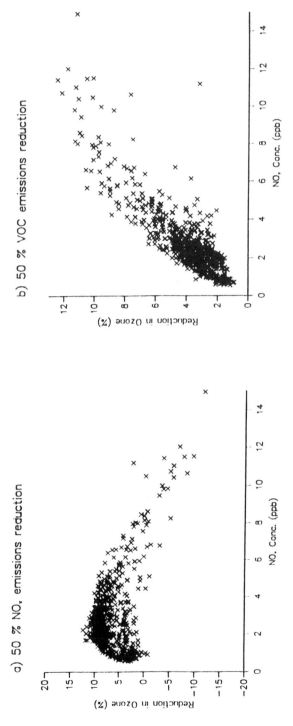

Fig. 3.14: (Left) Reductions in mean ozone concentration for April–September 1989 in each of 709 grid squares as a result of a 50 % VOC emission reduction, plotted against the NO$_x$ concentrations in the base case in each grid square. (Right) Reductions in mean ozone concentration for April–September 1989 in each of 709 grid squares as a result of a 50 % NO$_x$ emission reduction, plotted against the NO$_x$ concentrations in the base case in each grid square (from Simpson, 1995).

In the evaluation of emission reduction policies for hydrocarbons, the role of individual hydrocarbons in the formation of ozone is calculated by the so-called photochemical ozone creation potential (POCP), which expresses the ozone forming ability of any VOC relative to that of ethene (Derwent and Jenkin, 1990). Several studies have shown that POCP can be used to rank VOCs in terms of their importance to ozone formation. There are however, significant variabilities in the POCPs assigned to each VOC species (Andersson-Sköld *et al.*, 1991; Simpson, 1995), and these are determined both by how the value is recorded (as a peak value or as an average value), the NO_x level and by the physical conditions (cloudiness, temperature, solar zenith angle, *etc.*). During the chemical ageing of an air mass, NO_x has a faster loss rate than the VOCs and eventually the concentration of NO_x becomes so low that only a little ozone is produced, even though slowly reacting VOCs including CO are still present. The POCP parameter does not indicate that the ozone forming potential of a given volatile organic compound depends on the NO_x concentration.

c. Emission reductions required to achieve specific environmental policy goals

A direct policy application was made when the effect on ozone and other photo-oxidants of the specific VOC emissions reductions agreed in the UN-ECE VOC Protocol in 1991 was investigated in the EMEP project (Simpson and Styve, 1992). The VOC Protocol should result in a reduction of at least 15 % in European man-made VOC emissions compared to 1989. This was calculated to cause a reduction in mean ozone levels of around 4–8 % in north-western Europe and 1-4 % elsewhere in Europe. For ozone levels over 75 ppb, the excess was calculated to be reduced by 40–60 % in most areas. The costs or time required to implement these emission reductions were not evaluated; neither were the consequences for other environmental problems such as acidification and eutrophication, (see Grennfelt *et al.*, 1994). The model results both for precursors and secondary products have to some extent been evaluated with results from TOR. The finding of a possible NO_x limitation in the Freiburg urban plume shows a need for further work to strengthen the validity of model results for policy purposes.

d. Potential EUROTRAC contributions to environmental policy solutions for regional or continental photo-oxidants

The EURAD chemical transport model has been applied in calculating the horizontal transport of ozone and its precursors across regional boundaries in Europe (Memmesheimer, 1994), and there is considerable potential for further development in this direction. When calculations of regional fluxes of ozone and precursors from a chemical transport model like EURAD are used together with measured data sets both for chemistry and meteorology, complementary information is obtained about the formation of ozone from precursors emissions over different parts of Europe. In this way policy instruments can be developed which are based on the available information from field experiments, from theory and from laboratory studies of atmospheric chemical and physical processes.

The analysis of the TOR data will provide substantial scientific support for the negotiations in the second generation of abatement strategy protocols under the UN-ECE LRTAP Convention, in particular the revised NO_x protocol, as well for work within the European Environmental Agency (EEA) and the EU Framework Directive for Ozone. The contributions of central and eastern European TOR participants are important here. The measurements at TOR stations could eventually provide information about trends, both in the frequency of peak concentrations as well as in average levels, but long time series of observations are required and there is a need to extend the operation of the TOR network beyond 1995.

The collaboration between GENEMIS, the UN-ECE Task Force on Emissions and the EU CORINAIR project is expected to lead to a comprehensive emission inventory for Europe with a 50×50 km^2 resolution in space and 1 h in time. This would contribute to meeting the objectives of both the UN-ECE and the EEA. The link which is established between the emission data and economical sectors based on socio-economic and technical data could be used to prepare emission scenarios for alternative policies. The first efforts in this respect have been made in Austria (Friedrich *et al.*, 1994). The joint work of GENEMIS, BIATEX and EUMAC on an improved European emission inventory can lead to further model studies of how source-receptor relationships, established from biogenic and natural emissions, are changed by anthropogenic emissions.

The analysis of the exchange of ozone between the free troposphere and the atmospheric boundary layer can be further quantified through the use of data from TOR and the methods developed in EUMAC. The influence of the tropospheric concentrations on the boundary layer photo-oxidant distribution could then be better quantified. The models developed within EUROTRAC could assist in establishing source-receptor relationships between photo-oxidants concentrations and the precursor emissions inside and outside of Europe.

A possible contribution by EUROTRAC (subprojects EUMAC and TRACT) would be a quantification of how topography and land use influence source-receptor relationships for Europe. So far, the study of how complex terrain and land and sea surfaces influence heat exchange and air transport has mainly been confined to the local and sub-regional scales (see section 3.3.1 and Moussiopoulos, 1994a).

The EURAD model system is designed as a forecast tool which has a potential for predicting regional photo-smog episodes. It could contribute to the development of operational air pollution models by weather services. Such applications also need the values of the emissions which, in EURAD, were simulated by a specific emission model. The forecasting of emissions is a major goal of GENEMIS and BIATEX.

The models developed and used in EUMAC and TOR could also be applied to investigate the effect of alternative policy options for emission reduction on photo-oxidant peak and average concentration levels. Some of the environmental policy questions raised above could be elaborated by calculating the effect on the pollutant level of the maximum emission reductions which are achievable technically or economically, and if exceedances of standard or target levels still remain, those can be assessed. The degree of emission reduction needed to avoid the exceedance of standards or target levels could be calculated and include optimisation where the costs of the emission reductions for the various chemical components and in the different economical sectors would be taken into account.

3.3.3   Photo-oxidants on the global scale

The measurements and model calculations made in TOR and GLOMAC have contributed appreciably to a better understanding of the global ozone budget. This has important implications for environmental policy since it has been established that in the northern hemisphere about 50 % of the ozone content is related to anthropogenic precursor emissions, and the flux of ozone out of the North American continent in summer corresponds to about 30 % of the stratospheric flux over the northern hemisphere. On the other hand, when all direct and indirect effects of atmospheric ozone change are included, the climate impact is not well known. It has also been established that ozone from North America can be a source for ozone in the atmospheric boundary layer over Europe, but the magnitude is not established. The knowledge of the cycle of tropospheric ozone in the northern hemisphere is not sufficient at present to make reliable policy recommendations.

More specifically, the results from long-term ozone soundings made at Uccle (De Muer and De Backer, 1994) and the ground based measurements at the Zugspitze TOR station (Scheel et al., 1993; Sladkovic et al., 1993) have utilised by the WMO (1994) to identify and support the trend of ozone concentrations in the free troposphere of the northern hemisphere over the last two decades (see Fig. 3.1 and 3.2) and have also contributed to the evaluation of the secular increase in tropospheric ozone. It was shown that the concentration of ozone in the troposphere north of 20 °N has increased since the beginning of modern measurements in the 1970s (Fig. 3.2). This increase was larger at northern mid latitudes than in the tropics and larger over Europe and Japan than over North America. When compared with historical data it has been suggested that ozone in the troposphere over Europe has doubled since the turn of the century and that most of the increase has occurred since the 1950s. In the 1980s, the ozone increase was much smaller than in the 1970s (see section 3.2.1.).

Although not yet fully analysed, the data from the TOR station at Izaña, Tenerife support the conclusion that export of ozone and precursors from the North

American continent have an appreciable impact on ozone levels in the free troposphere over the Atlantic and Europe.

Long-term trends of greenhouse gases, including gases that contribute to atmospheric ozone chemistry, have been established from measurements at the Jungfraujoch. Accurate data for trends are a requirement when the validity of environmental policies is assessed. Long-term records are established of total column abundances of a number of trace gases, such as CH_4, N_2O, CO and CFCs (Fig. 3.15). These time series have been used in WMO (1994) and IPCC (1994) in trend assessment. Jungfraujoch has provided the observations based on infrared absorption for the primary Network for the detection of stratospheric change (NDSC) since 1991.

a. Potential for the application of EUROTRAC results for photo-oxidants on the global scale photo-oxidants on a global scale

While models have been used for assessment of the global ozone budget and its sensitivity to changes in precursor concentrations, the model comparison performed for IPCC (1994) and WMO (1994) identified large differences in the calculated results. Therefore such assessments do not have the confidence limits necessary for abatement strategies. The continued improvement in model formulation from EUMAC, GLOMAC, TOR, and TRACT in combination with the increased information that is being gained from atmospheric measurements will eventually lead to a significant improvement in understanding and quantifying the most important processes. Hence, it will eventually enable more accurate predictions of man's contribution to global change to be made. It was, however, noted by WMO in 1994 that a better understanding of the budget of ozone and OH on a global scale requires much better information on the concentrations of ozone, H_2O and, particularly NO_x and also on the source strength of NO_x from lightning.

Improvements could come from the EUROTRAC work listed below: a better understanding of the most important photolysis rates will come from measurements that are presently carried out in TOR (and from other European and non-European projects as a result of technology transfer from TOR). A better understanding of the distribution and seasonality of the concentrations of VOC and NO_y compounds will be derived from the analysis of the on-going measurements in TOR. TOR stations such as Izaña, the Jungfraujoch, the Zugspitze and, possibly, Schauinsland will contribute substantially to the Global Atmosphere Watch (GAW) network of WMO. The ozone sonde activities in TOR have been of use in other programmes, in particular in a number of EC projects such as TOASTE, OCTA and MOZAIC, and have led to improvements in the monitoring of the vertical profile of ozone. As a consequence, it is planned to extend the calibration facility for ozone sondes on the basis of the existing TOR ozone sonde calibration facility in Jülich (Smit *et al.*, 1994).

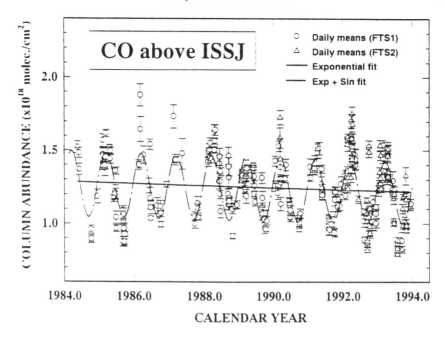

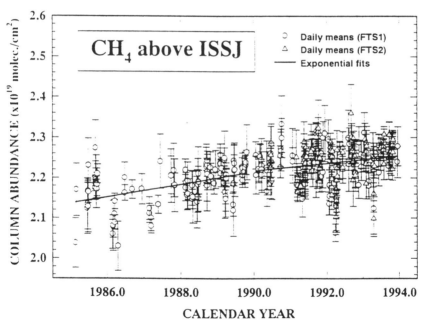

Fig. 3.15:   Trends in atmospheric abundances of CO and CH$_4$ measured at the TOR station, Jungfraujoch (Zander *et al.*, 1994a,b). Between 1984 and 1993 there was an exponential decrease in CO of (-0.55 ± 0.13) % yr^{-1} prior to 1990 to (0.41 ± 0.13) % yr^{-1} thereafter. For both molecules significant seasonal and sub-seasonal variability is superimposed on the mean changes.

3.4   References

Akimoto, H., Nakane, N. and Matsumoto, Y., 1994, The chemistry of oxidant generation: Tropospheric ozone increase in Japan, in: J.G. Calvert (ed), *The Chemistry of the Atmosphere: Its Impact on Global Change*, Blackwell Sci. Publ., pp. 261–273.

Alaart, M., Kelder, H. and Heijboer, L.C.G., 1994, On the transport of trace gases by extra-tropical cyclones, *Proc. Quadr. Ozone Symp.*, Charlottesville, June 4–13, 1992, in press.

Ambrosetti, P., Anfossi, D., Cieslik, S., Gaglione, P., Lamprecht, R., Mazorati, A., Sandroni, S., Spinedi, F., Stingele, A. and Vogt, S., 1991, The TRANSALP-89 Exercise. A tracer release experiment in a sub-alpine valley, *Ispra Joint Research Centre Report EUR13474 EN*, CEC.

Ambrosetti, P., Anfossi, D., Cieslik, S., Graziani, G., Grippa, G., Lamprecht, R., Mazorati, A., Stingele, A. and Vogt, S., 1994a, The TRANSALP 90 Campaign. A second tracer release experiment in a sub-alpine valley, *Ispra Joint Research Centre Report EUR15952 EN*, CEC.

Ambrosetti, P., Anfossi, D., Cieslik, S., Graziani, G., Grippa, G., Lamprecht, R., Mazorati, A., Nodop, K., Stingele, A. and Zimmermann, H., 1994b, Mesoscale transport of atmospheric trace gases across the Alps: the TRANSALP 1991 Campaign, *EUR16046 EN*, CEC.

Andersson-Sköld, Y., Grennfelt, P. and Pleijel, K., 1991, Photochemical ozone creation potential - a study of different concepts. Swedish Environmental Research Institute, *Report L91/248*.

Andreae, M. O., 1994, Climatic effects of changing atmospheric aerosol levels, in: A. Henderson-Sellers (ed), *World Survey of Climatology. Vol. **16:** Future Climates of the World*, Elsevier, Amsterdam, in press.

Anfossi,D., Sandroni, S. and Viarengo, S., 1991, Tropospheric ozone in the nineteenth century: The Moncalieri series. *J.Geophys.Res.* **96D**, 17349–17352.

Ashmore, M.R., Bell, J.N.B. and Brown, I.J., 1990, Air pollution and forest ecosystems in the European Community. *Air Pollution Research Report* **29**, Brussels.

Atkins, D.H.F., Cox, R.A. and Eggleton, A.E.J., 1972, Photochemical ozone and sulfuric acid formation in the atmosphere over southern England. *Nature* **235**, 372–376.

Atlas, E.L, 1988, Evidence for C_3–C_8 alkyl nitrates in rural and remote atmospheres, *Nature* **331**, 426–428.

Attmannspacher, W., Hartmannsgruber, R. and Lang, P., 1984, Langzeittendenzen des Ozons der Atmosphäre aufgrund der 1967 begonnenen Ozonmeßreihen am Meteorologischen Observatorium Hohenpeißenberg, *Meteorol. Rdsch.* **37**, 193–199.

Ayers, G.P., Penkett, S.A., Gillett, R.W., Bandy, B., Galbally, I.E., Meyer, C.P., Elsworth, C.M., Bentley, S.T. and Forgan, B.W., 1992, Evidence for photochemical control of ozone concentrations in unpolluted marine air, *Nature,* **360,** 446–448.

Baldasano J.M., Costa M., Cremades L., Flassak Th. and Wortmann-Vierthaler M., 1993, Influence of the traffic conditions on the air quality of Barcelona during the Olympic Games '92, *Proc. 20th ITM on Air Pollution Modelling and its Application*, pp. 513–514.

Barrie, L.A., J.W. Bottenheim, R.C. Schnell, P.J. Crutzen, and R.A. Rasmussen, 1988, Ozone destruction and photochemical reactions at polar sunrise in the lower Arctic atmosphere, *Nature* **334**, 138–140.

Beck, J.P. and Grennfelt, P., 1994, Estimate of ozone production and destruction over northwestern Europe. *Atmos. Environ.* **28**, 129–140.

Beck, J.P., Pul, W.A.J., Muer, D. De and Backer, H. De, 1994, Exchange of ozone between the atmospheric boundary layer and the free troposphere. *EUROTRAC Annual Report part 9: TOR,* EUROTRAC ISS, Garmisch-Partenkirchen, pp. 44–51.

Becker, K.-H., 1994, The atmospheric oxidation of aromatic hydrocarbons and its impact on photo-oxidant chemistry,in: P.M. Borrell, P. Borrell, T. Cvitaš, W. Seiler (eds), *Proc. EUROTRAC Symp. '94,* SPB Acedemic Publishing bv, The Hague 1994.

Beekmann, M., Ancellet, G. and Megie, G., 1994, Climatology of tropospheric ozone in southern Europe and its relation to potential vorticity, *J. Geophys. Res.,* **99**, 12,841–12,854.

Bonasoni, P., Fortezza, F., Georgiadis, T., Giovanelli, G. and Strocchi, V., 1991, Formation and transport of oxidants over a highly populated coastal area. *Chemrawn VII, Chemistry of the global atmosphere,* Baltimore 2–6 December 1991.

Borrego, C., Coutinho, M., and Barros, N., 1994a, Atmospheric pollution in the Lisbon airshed, in: Power, H., Moussiopoulos, N., and Brebbia, C. (eds), *Urban Air Pollution,* Computation mechanics Publications, **1** Chapter 3.

Borrego, C., Coutinho, M., and Barros, N., 1994b, Intercomparison of two meso-meteorological models applied to the Lisbon region, *Meteorology and App. Phys.* submitted.

Bouscaren, R., 1991, The problems related with the photochemical pollution in the Southern E.C. member states, *Final Report, Contract No 6611–31–89.*

Builtjes, P.J.H., 1985, The PHOXA-project, photochemical oxidants and acid deposition model application. *Proceedings of COST 611-workshop* at RIVM, The Netherlands, 23–25 September 1985.

Builtjes, P.J.H., 1992, The LOTOS-Long term ozone simulation project, summary report. TNO, P.O.Box 6011, NL-2600 Delft, The Netherlands.

Cerutti, C., Sandroni, S., Froussou, M., Asimakopoulos, D.N. and Helmis, C.G., 1989, Air quality in the Greater Athens Area. I. Monitoring campaign in September 1987. *Ispra Joint Research Centre Report EUR 12118 EN,* CEC.

Chin, M., Jacob, D.J., Munger, J.W., Parrish, D.D. and Doddridge, B.G., 1994, Relationship between ozone and carbon monoxide over North America, *J. Geophys. Res.* **99D**, 14565–14573.

Cox, R.A., Eggleton, A.E.J., Derwent, R.G., Lovelock, J.E. and Pack, D.H., 1975, Long range transport of photochemical ozone in northwestern Europe. *Nature* **255**, 118–121.

Crutzen, P.J., 1979, The role of NO and NO_2 in the chemistry of the troposphere and stratosphere. *Ann. Rev. Earth Planet. Sci. Publ.* **7**, 443–472.

Crutzen, P.J., 1988, Tropospheric ozone: An overview. In: I.S.A.Isaksen (ed), *Tropospheric Ozone,* Reidel, 3–32.

Cvitaš, T. and Kley, D., 1994, *The TOR network.* EUROTRAC ISS, Garmisch-Partenkirchen, 182 pp.

Danielsen, E. F. and Mohnen, V.A., 1977, Project Dustorm report: Ozone transport, in situ measurements and meteorological analyses of tropopause folding, *J. Geophys. Res.* **82**, 5867–5877.

De Muer, D. and Backer, H. De, 1994, Trend analysis of 25 of regular ozone soundings at Uccle (Belgium), in: P.M. Borrell, P. Borrell, T. Cvitaš, W. Seiler (eds), *Proc. EUROTRAC Symp. '92,* SPB Academic Publishing bv, The Hague 1993.

Dentener, F. J. and Crutzen, P.J., 1993, Reaction of N_2O_5 on tropospheric aerosols: Impact on global distributions of NO_x, O_3 and OH. *J. Geophys. Res.* **98**, 7149–7163.

Dentener, F. J., Lelieveld, J. and Crutzen, P.J., 1993, Heterogeneous reactions in clouds: consequences for the global budget of O_3. *Proc. CEC/EUROTRAC symposium in Varese, Italy*, 18–22 Oct.

Derwent, R.G. and Davies, T.J., 1994, Modelling the impact of NO_x or hydrocarbon control on photochemical ozone in Europe. *Atmos. Environ.* **28**, 2039–2052.

Derwent, R.G. and Hov, Ø., 1979, Computer modelling studies of photochemical air pollution formation in North West Europe. *AERE Report R-9434.* HMSO, London, pp. 147.

Derwent, R.G. and Jenkin, M., 1990, Hydrocarbon involvement in photochemical ozone formation in Europe. *AERE Report R-13736*, HMSO, London.

Derwent, R.G. and Kay, P.J.A., 1988, Factors influencing the ground level distribution of ozone in Europe. *Environ. Pollut.* **55**, 191–220.

Derwent, R.G., Simmonds, P.G. and Collins, W.J., 1994, Ozone and carbon monoxide measurements at a remote maritime location, Mace Head, Ireland from 1990–1992, *Atmos. Environ.* in press.

Dickerson *et al.*, 1987, Thunderstorms: An important mechanism in the transport of pollutants. *Science* **235**, 460–465.

Ebel, A., Becker, K:-H., Borrego, C.A., Bouscaren, R., Builtjes, P., Floßmann, A., Hansen, U., Hantel, M., Hass, H., Moussiopouöos, N., Poppe, D., Rosset, R., 1994, *EUROTRAC Annual Report part 5: EUMAC,* EUROTRAC ISS, Garmisch-Partenkirchen, pp. 1–8.

Ebel, A. and Petry, H., 1994, The impact of air traffic on tropospheric composition, in: Baldasano, J.M., Brebbia, C.A., Power, H., Zannetti, P. (eds), *Air Pollution II,* **1**, pp. 221–229.

Ebel, A., Elbern, H. and Oberreuter, A., 1993, Stratosphere-troposphere air mass exchange and cross-tropopause fluxes of ozone, in: Thrane, E.V., *et al.* (eds), *Coupling processes in the Lower and Middle Atmosphere,* Kluwer Academic Publishing, pp. 49–65.

Ebel, A., Elbern, H., Hendricks, J.and Meyer, R., 1995, Stratosphere-troposphere exchange and its impact on the structure of the lower stratosphere, *J. Geomag. and Geo-elec.* in press.

Ebel, A., Haas, H., Jakobs, H.J., Laube, M., Memmesheimer, M., Oberreuter, A., Geiß, H. and Kuo, Y-H., 1991, Simulation of the ozone intrusion caused by a tropopause fold and cut-off low, *Atmos. Environ.* **25A**, 2131–2144.

Ebel, A., Memmesheimer, M., Hass, H. and Jakobs, H.J., 1993, Evaluation of emission data by long range transport modelling in Europe (VALIMATRA). *EUROTRAC Annual Report part 5: GENEMIS,* EUROTRAC ISS, Garmisch-Partenkirchen, pp. 122–124.

Ehhalt, D.H., Rohrer, F. and Wahner, A., 1992, Sources and distribution of NO_2 in the upper troposphere at northern mid-latitudes, *J. Geophys. Res.* **97**, 3725–3738.

Eliassen, A., Hov, Ø., Isaksen, I.S.A. Saltbones, J. and Stordal, F., 1982, A Lagrangian long-range transport model with atmospheric boundary layer chemistry. *J. App. Meteorology* **21**, 1645–1661.

Fehsenfeld, F., Calvert, J., Fall, R., Goldan, P., Guenther, A.B., Hewitt, C.N., Lamb, B., Liu, S., Trainer, M., Westberg H. and Zimmerman, P., 1992, Emissions of volatile organic compounds from vegetation and their implications for atmospheric chemistry. *Global Biogeochemical Cycles* **6,** 389–430.

Feister, U. and Warmbt, W., 1987, Long-term measurements of surface ozone in the German Democratic Republic, *J. Atmos. Chem.* **5**, 121.

Fiedler, F., 1992, Das Regio-Klima-Projekt: Wie regeln die natuerlichen Energieumsetzungen das Klima in einer Region? *KfK Nachrichten* **24**, Kernforschungszentrum Karlsruhe, 125–131.

Finlayson-Pitts, B. J., Ezell, M.J. and Pitts, J.N.,Jr., 1989, Formation of chemically active chlorine compounds by reactions of atmospheric NaCl particles with gaseous N_2O_5 and $ClONO_2$, *Nature* **337**, 241–244.

Fishman, J., Brackett, V.G. and Fakhruzzaman, K., 1992, Distribution of tropospheric ozone in the tropics from satellite and ozonesonde measurements, *J. Atmos. Terrestrial Phys.* **54**, 589–597.

Flatøy, F., and Hov, Ø., 1995, 3-D model studies of the effect of NO_x emissions from aircraft on ozone in the upper troposphere over Europe and the North Atlantic, *J. Geophys. Res.* submitted.

Flatøy, F., Hov, Ø. and Smit, H., 1995, 3-D model studies of vertical exchange processes of ozone in the troposphere over Europe. *J. Geophys. Res.* in press.

Flocke, F., 1992, Messungen von Alkylnitraten (C_1–C_8) am Schauinsland im Schwarzwald. Ein Beitrag zur Bilanzierung der Photochemischem Ozonproduktion, *Inaugural-Dissertation zur Erlangung des Doktorgrades des Fachbereichs Naturwissenschaften II der Bergischen Universität-Gesamthochschule Wuppertal.*

Flocke, F., Gilge, S., Mihelcic, D., Smit, H.G.J., Volz-Thomas, A., Buers, H.J., Garthe, H.J., Geiss, H., Heil, T., Heitlinger, M., Herrmanns, B., Houben, N., Klemp, D., Kramp, F., Müsgen, P., Pätz, H.W., Schultz, M., Su, Y. and Kley, D., 1994, Photo-oxidants and precursors at Schauinsland, Black Forest: Results from continuous measurements of H_2O_2 and organic hydroperoxides and from in situ measurements of peroxy radicals by MIESR. *EUROTRAC Annual Report part 9: TOR,* EUROTRAC ISS, Garmisch-Partenkirchen, pp.166–176.

Flocke, F., Volz-Thomas, A. and Kley, D., 1994, The use of alkyl nitrate measurements for the characterization of the ozone balance at TOR-Station No. 11, Schauinsland, A contribution to subproject TOR, in: P.M. Borrell, P. Borrell, T. Cvitaš, W. Seiler (eds), *Proc. EUROTRAC Symp. '94*, SPB Academic Publishing bv, The Hague 1994, pp. 243–247.

Fox, C.B., 1873, *Ozone and Antozone* Churchill Publ., London.

Friedrich, R., Heymann, M. and Kasas, Y., 1994, The GENEMIS inventory: The estimation of European emission data with high temporal resolution, *EUROTRAC Annual Report part 5: GENEMIS,* EUROTRAC ISS, Garmisch-Partenkirchen, pp.1–14.

Gaeb, S., Brockmann, K.J., Rupert, L. and Turner, W.V., 1994, Formation of alkyl and 1-hydroxyalkyl hydroperoxides on ozonolysis in water and in air, *EUROTRAC Annual Report part 8: LACTOZ,* EUROTRAC ISS, Garmisch-Partenkirchen, pp.116–123.

Garland, J. and Derwent, R.G., 1979, Destruction at the ground and the diurnal cycle of concentration of ozone and other gases. *Quart. J. Roy. Met. Soc.* **105**, 169–183.

Geiß, H., Hass H., Jakobs H. and Memmesheimer M., 1994, Flux calculations of ozone and its precursors with the EURAD model at the TOR station Schauinsland, in: P.M. Borrell, P. Borrell, T. Cvitaš, W. Seiler (eds), *Proc. EUROTRAC Symp. '94,* SPB Academic Publishing bv, The Hague 1994.

Gidel, L.T. and Shapiro, M.A., 1980, General circulation model estimates of the net vertical flux of ozone in the lower stratosphere and the implications for the tropospheric ozone budget, *J. Geophys. Res.* **85**, 4049–4058.

Grennfelt, P., Hov, Ø. and Derwent, R.G., 1994, Second generation abatement strategies for NO_x, NH_3, SO_2 and VOC. *Ambio* **23,** 425–433.

Grennfelt, P., Saltbones, J. and Schjoldager, J., 1987, *Oxidant data collection in OECD-Europe 1985-87 (OXIDATE)*. April-September 1985. NILU OR 22/87, NILU, Lillestrøm, Norway.

Grennfelt, P., Saltbones, J. and Schjoldager, J., 1988, *Oxidant data collection in OECD-Europe 1985–87 (OXIDATE)*. *Report on ozone, nitrogen dioxide and peroxyacetyl nitrate* October 1985–March 1986 and April–September 1986. NILU OR 31/88, NILU, Lillestrøm, Norway.

Guicherit, R., 1988, Ozone on an urban and regional scale - with special reference to the situation in the Netherlands, in: I.S.A. Isaksen (ed), *Tropospheric Ozone*, Reidel Publ., pp. 49–62.

Güsten, H., Heinrich, G., Mönnich, E., Sprung, D. and Weppner, J., 1993, TRACT'92 field measurement campaign in the nested area - Preliminary results. *EUROTRAC Annual Report: ALPTRAC and TRACT,* EUROTRAC ISS, Garmisch-Partenkirchen, pp. 82–90.

Haagen-Smit, A.J., 1952, Chemistry and physiology of Los Angeles smog, *Indust. Eng. Chem.* **44,** 1342–1346.

Hansen, U., Adolph, D. and Kung, J.G., 1994, Update and temporal resolution of emissions from large point sources in GENEMIS. *EUROTRAC Annual Report part 5: GENEMIS,* EUROTRAC ISS, Garmisch-Partenkirchen, pp.43–62.

Hass, H., Ebel, A., Jakobs, H.J. and Memmesheimer, M., 1993, Interaction of the dynamics and chemistry in photo-oxidant formation, in: P.M. Borrell, P. Borrell, T. Cvitaš, W. Seiler (eds), *Proc. EUROTRAC Symp. '94*, SPB Academic Publishing bv, The Hague 1994, pp. 65–68.

Hayman, G.D., 1994, The oxidation of biogenic hydrocarbons. in: P.M. Borrell, P. Borrell, T. Cvitaš, W. Seiler (eds), *Proc. EUROTRAC Symp. '94*, SPB Academic Publishing bv, The Hague 1994, pp. 75–82.

Helas, G., 1994, The shopping list of GENEMIS to BIATEX. *EUROTRAC Annual Report part 4: BIATEX,* EUROTRAC ISS, Garmisch-Partenkirchen, pp. 240–251.

Hewitt, C.N. and Street, R.A., 1992, A qualitative assessment of the emission of non-methane hydrocarbon compounds from the biosphere to the atmosphere in the UK: present knowledge and uncertainties. *Atmos. Environ.* **26A,** 3069–3077.

Heymann, M., 1993, The temporal resolution of emissions in Europe. In G.McInnes, J.M.Pacyna and H.Dovland (eds), *Proc. 2nd meeting of the Task Force for Emission Inventories*, Delft EMEP/CCC Report 8/93, NILU, P.O.Box 100, N-2007 Kjeller, Norway, pp. 281–292.

Hov, Ø and Derwent, R.G., 1981, Sensitivity studies of the effects of model formulation on the evaluation of control strategies for photochemical air pollution formation in the United Kingdom. *J. Air Pollution Control Association* **12,** 1260–1267.

Hov, Ø. and Stordal, F., 1993, Measurements of ozone and precursors at Ny Aalesund on Svalbard and Birkenes on the south coast of Norway, ozone profiles at Bjørnøya, and interpretation of measured concentrations, 1992. *EUROTRAC Annual Report part 9 TOR,* EUROTRAC ISS, Garmisch-Partenkirchen, pp. 175–183.

Hov, Ø., 1985, The effect of chlorine on the formation of photochemical oxidants in Southern Telemark. *Atmos. Environ.* **19,** 471–485.

Hov, Ø., 1989, Changes in Tropospheric Ozone: A Simple Model Experiment, in: R.D. Bojkov, P.Fabian (eds), *Ozone in the Troposphere*, Deepak Publ.

Hov, Ø., Hesstvedt, E. and Isaksen, I.S.A., 1978, Long-range transport of tropospheric ozone. *Nature* **242**, 341–344.

Hov, Ø., Stordal, F. and Eliassen, A., 1985, Photochemical oxidant control strategies in Europe: A 19 days' case study using a Lagrangian model with chemistry. NILU TR 5/85. NILU, P.O.Box 100, N-2007 Kjeller, Norway.

IPCC, 1994, Climate Change 1994, *The IPCC Scientific Asssessment,* Intergovernmental Panel on Climate Change, Cambridge University Press

Isaksen, I.S.A. and Hov, Ø., 1987, Calculation of trends in the tropospheric concentrations of O_3, OH, CO, CH_4 and NO_x. *Tellus* **39B**, 271–285.

Jacob, D., Logan, J.A., Gardener, G.M., Yevich, R.M., Spivakowsky, C.M. and Wofsy, S.C., 1993, Factors regulating ozone over the United States and its export to the global atmosphere, *J. Geophys. Res.* **98**, 14817–14826.

Jaeschke, W., Dietrich, Th. and Schickedanz, U., 1994, Determination of CO_2, NO, NO_2 and O_3 fluxes between the atmosphere and a strongly polluted forest ecosystem, *EUROTRAC Annual Report part 4: BIATEX,* EUROTRAC ISS, Garmisch-Partenkirchen, pp. 102–108.

Jakobs, H., 1994, Use of nested models for air pollution studies. in: P.M. Borrell, P. Borrell, T. Cvitaš, W. Seiler (eds), *Proc. EUROTRAC Symp. '94,* SPB Academic Publishing bv, The Hague 1994, p. 799.

Jonson, J. E., and Isaksen, I.S.A., 1993, Tropospheric ozone chemistry: The impact of cloud chemistry, *J. Atmos. Chem,* **16**, 99–122.

Kalthoff, N., Binder, H.-J., Fiedler, F., Kosmann, M., Corsmeier, U., Voegtlin, R. and Schlager, H., 1994, The boundary layer evolution during TRACT, in: P.M. Borrell, P. Borrell, T. Cvitaš, W. Seiler (eds), *Proc. EUROTRAC Symp. '94,* SPB Acedemic Publishing bv, The Hague 1994, pp. 748–752.

Kible, R. and Smiatek, G., 1994, Mapping land use for modeling emission and deposition in Europe. *EUROTRAC Annual Report part 5: GENEMIS,* EUROTRAC ISS, Garmisch-Partenkirchen, pp. 74–79.

Klemp, D., Flocke, F., Kramp, F., Pätz, W., Volz-Thomas, A. and Kley, D., 1993, Indications for biogenic sources of light olefins in the vicinity of Schauinsland, Black Forest (TOR station no. 11). In: J. Slanina, G Angeletti and S. Beilke (eds), *Air Pollution Research Report* **47** CEC, Brussels, pp. 271–281.

Kley, D., Volz, A. and Mülhaus, F., 1988, Ozone measurements in historic perspective. In: I.S.A.Isaksen (ed), *Tropospheric Ozone*, Reidel Publ., pp. 63–72.

Kramp, F., Buers, H.J., Flocke, F., Klemp, D., Kley, D., Pätz, H.W., Schmitz, T., Volz-Thomas, A., Determination of OH-concentrations from the decay of C_5–C_8 hydrocarbons between Freiburg and Schauinsland: Implications on the budgets of olefins, a contribution to subproject TOR, in: P.M. Borrell, P. Borrell, T. Cvitaš, W. Seiler (eds), *Proc. EUROTRAC Symp. '94,* SPB Academic Publishing bv, The Hague 1994, pp. 373–378.

Lalas, D.P., Asimakopoulos, D.N., Deligiorgi, D.G. and Helmis, C.G., 1983, Sea breeze circulation and photochemical pollution in Athens, Greece, *Atmos. Environ.* **17**, 1621–1632.

Lelieveld, J. and Crutzen, P.J., 1994, Role of deep cloud convection in the ozone budget of the troposphere, *Science* **264**, 1759–1761.

Lelieveld, J., and Crutzen, P.J., 1990, Influences of cloud photochemical processes on tropospheric ozone, *Nature* **343**, 227–233.

Levy, H.B., Mahlman, J.D., Moxim, W.J. and Liu, S., 1985, Tropospheric ozone: The role of transport, *J. Geophys. Res.* **90**, 3753–3771.

Lin, X., Trainer, M. and Liu, S.C., 1988, On the nonlinearity of the tropospheric ozone production, *J. Geophys. Res.* **93**, 15879–15888.

Linvill, D.E., Hooken, W.J. and Olson, B., 1980, Ozone in Michigan's environment (1876–1880), *Mon. Weather Rev.* **108**, 1883–1891.

Liu, S. C., M. Trainer, F. C. Fehsenfeld, D. D. Parrish, E. J. Williams, D. W. Fahey, G. Hubler, and P. C. Murphy, 1987, Ozone production in the rural troposphere and the implications for regional and global ozone distributions, *J. Geophys. Res.* **92**, 4191–4207.

Liu, S.C., M. McFarland, D. Kley, O. Zafiriou, and B. Huebert, 1983, Tropospheric NO_x and O_3 budgets in the Equatorial Pacific, *J. Geophys. Res.* **88**, 1349–1368.

Logan, J.A., 1994, Trends in the vertical distribution of ozone: An analysis of ozonesonde data. *J. Geophys. Res.* submitted.

London, J. and S. Liu, 1992, Long-term tropospheric and lower stratospheric ozone variations from ozonesondes observations, *J. Atmos. Terr. Phys.* **5**, 599–625.

Low, P.S., P.S. Kelly and T.D. Davies, 1992, Variations in surface ozone trends over Europe, *Geophys. Res. Let.* **19**, 1117–1120.

Lübkert, B. and Schöpp, W., 1989, A model to calculate natural VOC emissions from forests in Europe, IIASA, *WP-89-082.*

Mak, J.E., 1992, Evidence for a missing carbon monoxide sink based on tropospheric measurements of ^{14}CO, *Geophys. Res. Lett.* **19**, 1467–1470.

Marenco, A., Philippe, N. and Hervé, G., 1994, Ozone measurements at Pic du Midi observatory. *EUROTRAC Annual Report part 9: TOR,* EUROTRAC ISS, Garmisch-Partenkirchen, pp.121–130.

Memmesheimer, M., 1994, TOR Task Group 4 status report. *EUROTRAC Annual Report part 9: TOR,* EUROTRAC ISS, Garmisch-Partenkirchen, pp. 52–62.

Millán, M., 1992, Report to the Commission of the European Communities on the MECAPIP project. In preparation, Centro de Estudios Ambeliantes del Mediterraneo (CEAM), Valencia, Spain.

Millán, M., 1994, Regional processes and long range transport of air pollutants in southern Europe. The experimental evidence. In G.Angeletti and G.Restelli (eds), *Physico-chemical behaviour of atmospheric pollutants, Proc. 6th European symp.*, *Air Pollution Research Report* **50**, EC, pp. 447–459.

Miller, A.J., Tiao, G.C., Reinsel, G.C., Wuebbles, D., Bishop, L., Kerr, J., Nagatani, R.M. and DeLuisi, J.J., 1994, Comparisons of observed ozone trends in the stratosphere through examination of Umkehr and balloon ozonesonde data, in preparation.

Moussiopoulos N., (ed), 1994a, *The EUMAC Zooming Model,* EUROTRAC ISS, Garmisch-Partenkirchen.

Moussiopoulos N., 1994b, Air pollution in Athens, in N. Moussiopoulos, C. Brebbia, H. Power (eds), *Urban Air Pollution,* Computational Mechanics Publications, in press.

Moussiopoulos, N. and Sahm, P., 1994, Further development and application of the EUMAC Zooming Model, *EUROTRAC Annual Report part 5: EUMAC,* EUROTRAC ISS, Garmisch-Partenkirchen, pp.80–85.

Murphy, D. M., and Fahey, D.W., 1994, An estimate of the flux of stratospheric reactive nitrogen and ozone into the troposphere, *J. Geophys. Res.* **99**, 5325–5332.

Nester, K., Fiedler, F. and Panitz, H.J., 1993, Coupling of the DRAIS model to the EURAD model and the analysis od sub-scale phenomena, *EUROTRAC Annual Report part 5: EUMAC,* EUROTRAC ISS, Garmisch-Partenkirchen, pp. 86–91.

Oke, T.R., 1987, *Boundary Layer Climates,* Routledge, London, 435 pp.

Oltmans, S. J. and Levy, H.II, 1994, Surface ozone measurements from a global network, *Atmos Environ.* **28,** 9–24.

Palmer F.H., 1994, European challenges for cleaner road transport fuels and engines, Seminar on *Urban Air Quality - Environmental and Health Effects of Motor Vehicle Fuel Composition,* Athens, 27 May 1994.

Pankrath, J., 1989, Photochemical oxidant model application within the framework of control strategy development in the Dutch/German programme PHOXA. In: Schneider, T., S.D.Lee, G.J.R.Wolters and L.D.Grant (eds), *Atmospheric ozone research and its policy implications, Proc. 3rd US-Dutch International Symp.,* Elsevier, Amsterdam, pp. 633–646.

Parrish, D.D., Holloway, J.S., Trainer, M., Murphy, P.C., Forbes, G.L. and Fehsenfeld, F.C., 1993, Export of North American ozone pollution to the North Atlantic Ocean, *Science* **259,** 1436–1439.

Penkett, S.A., Bandy, B.J., Burgess, R.A., Clemitshaw, K.C., Cardenas, L. and Carpenter, L., 1994, Measurements of NO_x, NO_y, CO an ozone at Weybourne and Mace Head. *EUROTRAC Annual Report part 9: TOR,* EUROTRAC ISS, Garmisch-Partenkirchen, pp. 275–284.

Penkett, S.A., Blake, N.J., Lightman, P., Marsh, A.R.W., Anwyl, P. and Butcher, G., 1993, The seasonal variation of nonmethane hydrocarbons in the free troposphere over the North Atlantic Ocean: Possible evidence for extensive reaction of hydrocarbons with the nitrate radical, *J. Geophys. Res,* **98,** 2865–2885.

Petry, H., Elbern, H., Lippert, E. and Meyer, R., 1994, Three dimensional mesoscale simulations of aeroplane exhaust impact in a flight corridor, in *Proc. Internat. Colloq. on the Impact of Emissions from Aircraft and Spacecraft upon the atmosphere, Cologne,* April 1994, DLR-Mitteilung 94–06, pp. 229–335.

Platt, U., Perner, D., Schröder, J., Kessler, C. and Toenissen, A., 1981, The diurnal variation of NO_3, *J. Geophys. Res.* **86,** 11965–11970.

Platt, U., Rateike, M., Junkermann, W., Rudolph, J. and Ehhalt, D.H., 1988, New tropospheric OH measurements, *J. Geophys. Res.* **93,** 5159–5166.

Poppe, D., Kuhn, M., Zimmermann, J., Koppmann, R. and Rohrer, F., 1994, The role of isoprene in the formation of ozone in the planetary boundary layer during a summer smog event, *EUROTRAC Annual Report part 5: EUMAC,* EUROTRAC ISS, Garmisch-Partenkirchen, pp.97–101.

PORG, 1993, Ozone in the United Kingdom 1993. *Third report of the United Kingdom Photochemical Oxidants Review Group,* Dep. of Environment, London, 170 pp.

Prevot, A.S.H., Staehelin, J., Brunner, D., Hering, A., Neininger, B. and Fust, P., 1994, Photo-oxidants in the southern pre-Alpine region of Switzerland and the northern part of Italy. Discussion of field measurements from summer 1992. In G.Angeletti and G.Restelli (eds), *Physico-chemical behaviour of atmospheric pollutants, Proc. 6th European symposium, Air Pollution Research Report* **50,** EC, pp. 519–524.

Pszenny, A.A.P., Keene, W.C., Jacob, D.J., Fan, S., Maben, J.R., Zetwo, M.P., Springer-Young, M. and Galloway, J.N., 1993, Evidence of inorganic gases other than hydrogen chloride in marine surface air, *Geophys. Res. Lett,* **20,** 699–702.

Ridley, B. A., Madronich, S., Chatfield, R.B., Walega, J.G., Shetter, R.E., Carroll, M.A. and Montzka, D.D., 1992, Measurements and model simulations of the photostationary state during the Mauna Loa Observatory Photochemistry Experiment: Implications for radical concentrations and ozone production and loss rates, *J. Geophys. Res.* **97**, 10375–10388.

Sandroni, S., Anfossi, D. and Viarengo, S., 1992, Surface ozone levels at the end of the ninteenth Century in South America, *J. Geophys. Res.* **97**, 2535–2540.

Sandroni, S., Bacci, P., Boffa, G., Pellegrini, U. and Ventura, A., 1994, Tropospheric ozone in the pre-alpine and alpine regions, *Science of the Total Environ.*, **156**, 169–182.

Schajor, R., Preunkert, S., Herbstreit, K. and Wagenbach, D., 1994, Deposition of heavy metals on to high elevation alpine snow fields, in: P.M. Borrell, P. Borrell, T. Cvitaš, W. Seiler (eds), *Proc. EUROTRAC Symp. '94*, SPB Academic Publishing bv, The Hague 1994, p. 735.

Scheel, H.E., Brunke, E.G. and Seiler, W., 1990, Trace Gas Measurements at the Monitoring Station Cape point, South Africa, Between 1978 and 1988, *J. Atmos. Chem*, **11**, 197–210.

Scheel, H.E., R. Sladkovic and W. Seiler, 1994, Ground-based measurements of ozone and related precursors at 47°N, 11°E, *EUROTRAC Annual Report part 9: TOR*, EUROTRAC ISS, Garmisch-Partenkirchen.

Scheel, H.E., Sladkovic, R. and Seiler, W., 1993, Ground-Based measurements of ozone and related precursors at 47°N, 11°E, in: P.M. Borrell, P. Borrell, T. Cvitaš, W. Seiler (eds), *Proc. EUROTRAC Symp. '92*, SPB Academic Publishing bv, The Hague 1993, pp. 104–108.

Schmitt, R. and Hansen, L., 1993, Ozone in the free troposphere over the north Atlantic: Production and long range transport. *EUROTRAC Annual Report part 9: TOR*, EUROTRAC ISS, Garmisch-Partenkirchen, pp. 112–118.

Schmitt, R., 1994, Ozone in the free troposphere over the North Atlantic: Production and Long-range Transport, *EUROTRAC Annual Report part 9: TOR*, EUROTRAC ISS, Garmisch-Partenkirchen, pp. 162–165.

Schmitt, R., Matuska, P., Carretero, P., Hanson, L. and Thomas, *K., 1994, Ozon in der freien Troposphäre: Produktion und großräumiger Transport.* Abschlußbericht des Vorhabens 07 EU 764, Bundes minister für Forschung und Technologie, Meteorologie Consult GmbH, 71 pp. Glashütten.

Schmitt, R., P. Matuska, P. Carretero, L. Hanson and K. Thomas, 1993, *Ozon in der freien Troposphäre: Produktion und großräumiger Transport*, Abschlußbericht zum Vorhaben 07 EU 764, Bundes minister für Forschung und Technologie, Meteorologie Consult GmbH, 71 pp. Glashütten.

Schneider, Ch., 1994, Emissionskataster zur Berechnung der Ausbreitung von Luftschadstoffen, PhD Dissertation, Universität Karlsruhe.

Schumann, U., 1994, On the effect of emissions from aircraft engines on the state of the atmosphere, *Ann. Geophys*, **12**, 365–384.

Seiler, W. and Fishman, J., 1981, The distribution of carbon monoxide and ozone in the free troposphere, *J. Geophys. Res*, **86**, 7255–7265.

Sigg, A. and Neftel, A., 1991, *Nature* **351**, 557.

Simmonds, P.G., 1993, Tropospheric ozone research and global atmospheric gases experiment, Mace Head, Ireland. *EUROTRAC Annual Report part 9: TOR*, EUROTRAC ISS, Garmisch-Partenkirchen, pp. 234–242.

Simpson, D. and Styve, H., 1992, The effects of the VOC protocol on ozone concentrations in Europe. *EMEP MSC-W Note 4/92*, The Norwegian Meteorological Institute, P.O.Box 43, Blindern, N-0313 Oslo, 24 pp.

Simpson, D., 1991, Long period modelling of photochemical oxidants in Europe. Calculations for April–September 1985, April–October 1989. *EMEP MSCW Report 2/91*. The Norwegian Meteorological Institute, P.O.Box 43-Blindern, N-0313 Oslo, Norway.

Simpson, D., 1995, *J. Geophys. Res.* **100 D11**, 22891–22906.

Simpson, D., 1993, Photochemical model calculations over Europe for two extended summer periods: 1985 and 1989. Model results and comparisons with observations, *Atmos. Environ,* **27A**, 921-943.

Simpson, D., 1994a, Biogenic VOC emissions in Europe. Part II: Implications for ozone strategies. *J.Geophys.Res.* submitted.

Simpson, D., 1994b, Biogenic VOC emissions in Europe Part I: Emissions and uncertainties. *EMEP/MSC-W note 5/94*, The Norwegian Meteorological Institute, P.O.Box 43 Blindern, N-0313 Oslo, Norway, 38 pp.

Singh, H.B., 1977, Preliminary Estimation of Average Tropospheric OH Concentrations in the Northern and Southern Hemispheres, *Geophys. Res. Lett.* **4,** 453–456.

Sladkovic, R., Scheel, H.E. and Seiler, W., 1993, Ozone climatology of the mountain sites, Wank and Zugspitze, in: P.M. Borrell, P. Borrell, T. Cvitaš, W. Seiler (eds), *Proc. EUROTRAC Symp. '92*, SPB Academic Publishing bv, The Hague 1994, pp. 104–108.

Slanina, J., Duyzer, J.H., Fowler, D., Helas, G., Hov, Ø., Meixner, F.X., Struwe, S., 1994, *EUROTRAC Annual Report part 4: BIATEX*, EUROTRAC ISS, Garmisch-Partenkirchen, pp. 1–33.

Smit, H.G.J., D. Kley, H. Loup, and W. Sträter, 1993, Distribution of Ozone and Water Vapour Obtained from Soundings over Jülich: Transport versus Photo-Oxidants, in: P.M. Borrell, P. Borrell, T. Cvitaš, W. Seiler (eds), *Proc. EUROTRAC Symp. '92*, SPB Academic Publishing bv, The Hague 1993, pp. 143–148.

Smit, H.G.J., Sträter, W., Kley, D., Proffitt, M.H., 1994, The evaluation of ECC ozone sondes under quasi flight conditions in the Environmental simulation chamber at Jülich, in: P.M. Borrell, P. Borrell, T. Cvitaš, W. Seiler (eds), *Proc. EUROTRAC Symp. '94*, SPB Academic Publishing bv, The Hague 1994, pp. 349–353.

Smit, H.G.J.and D. Kley, 1993, Vertical Distribution of Toposheric ozone Ozone and its Correlation with Water Vapour Over the Equatorial Pacific ocean between 160E and 160W, *EOS 74 (43)*, 115.

Solberg, S., Hermansen, O., Joranger, E., Pedersen, U., Stordal, F., Tørseth, K. and Hov, Ø., 1994a, Tropospheric ozone depletion in the Arctic during spring. *NILU OR 27/94*. NILU, P.O.Box 100, N-2007 Kjeller, 47 pp.

Solberg, S., Stordal, F., Schmidbauer, N. and Hov, Ø., 1994b, Non-methane hydrocarbons (NMHC) at Birkenes in South Norway, 1988-1993. *NILU Report 47/93*.

Spivakovsky, C.M. *et al.*, 1990, Tropospheric OH in a three dimensional chemical tracer model: An assessment based on observations of CH_3CCl_3, *J.Geophys. Res.* **95**, 18441–18471.

Staehelin, J. and Schläpfer, K., 1994, Determination of VOC emission factors for road traffic with tunnel measurements. *EUROTRAC Annual Report part 5: GENEMIS*, EUROTRAC ISS, Garmisch-Partenkirchen, pp. 83–87.

Staehelin, J., Thudium, J., Buehler, R., Volz-Thomas, A. and Graber, W., 1994, Trends in surface ozone concentrations at Arosa (Switzerland), *Atmos. Environ.* **28**, 75–87.

Steinbrecher, R. and Rabong, R., 1994, Isoprene emissions rates of Norway spruce; climate chamber studies and field experiments. *EUROTRAC Annual Report part 4: BIATEX,* EUROTRAC ISS, Garmisch-Partenkirchen, pp. 162–171.

Steinbrecher, R., Junkermann, W., Steinbrecher, J. and Slemr, J., 1994, VOCs and peroxides in ambient air above an oak/pine forest in the Mediterranean region. *EUROTRAC Annual Report part 4: BIATEX,* EUROTRAC ISS, Garmisch-Partenkirchen, pp. 153–161.

Strand, A. and Hov, Ø., 1995, The impact of man-made and natural NO_x emissions on the upper tropospheric ozone, *Atmos. Environ.* in press.

Strand, A., and Hov, Ø., 1994, Two-dimensional global study of the tropospheric ozone production. *J. Geophys. Res.* **99D**, 22877–22895.

Thompson, A.M., and Cicerone, R.J., 1986, Possible perturbations to atmospheric CO, CH_4, and OH, *J. Geophys. Res.* **91,** 10853–10864.

Veldt, C., 1989, Leaf biomass data for the estimation of biogenic VOC emissions. *MT-TNO Report 89-306.*

Vögtlin, R., Loeffler-Mang, M., Kossmann, M., Corsmeier, U., Fiedler, F., Klemm, O. and Schlager, H., 1994, Spatial and temporal distribution of air pollutants over complex terrain during TRACT, in: P.M. Borrell, P. Borrell, T. Cvitaš, W. Seiler (eds), *Proc. EUROTRAC Symp. '94,* SPB Academic Publishing bv, The Hague 1993, pp. 770–775.

Volz, A. and Kley, D., 1988, Evaluation of the Montsouris series of ozone measurements made in the nineteenth century. *Nature* **332,** 240–242.

Volz, A., Ehhalt, D.H. and Derwent, R.G., 1981, Seasonal and Longitudinal Variation of ^{14}CO and the Tropospheric Concentration of OH Radicals. *J. Geophys. Res.* **86(C6),** 5163–5171.

Volz-Thomas, A., 1993, Trends in photo-oxidant concentrations, in: P.M. Borrell, P. Borrell, T. Cvitaš, W. Seiler (eds), *Proc. EUROTRAC Symp. '92,* SPB Academic Publishing bv, The Hague 1993, pp. 59–64.

Volz-Thomas, A., Flocke, F., Garthe, H.J., Geiß, H., Gilge, S., Heil, T., Kley, D., Klemp, D., Kramp, F., Mihelcic, D., Pätz, H.W., Schultz, M. and Su, Y., 1993, Photo-oxidants and Precursors at Schauinsland, Black Forest, in: P.M. Borrell, P. Borrell, T. Cvitaš, W. Seiler (eds), *Proc. EUROTRAC Symp. '92,* SPB Academic Publishing bv, The Hague 1993, pp. 98–103.

Wagenbach, D., Jung, W., Schajor, R. and Preunkert, S., Retrospective and present state of anthropogenic aerosol deposition at a high altitude alpine glacier (Colle Gnifetti, 4450 m a.s.l.), *EUROTRAC Annual Report part 3: ALPTRAC,* EUROTRAC ISS, Garmisch-Partenkirchen, pp. 66–74.

Wayne, R.P., Barnes, I., Biggs, P., Burrows, J.P., Canosa-Mas, C.E., Hjorth, J., LeBras, G., Moortgat, G.K., Perner, D., Poulet, G., Restelli, G. and Sidebottom, H., 1991, The nitrate radical: Physics, chemistry, and the atmosphere 1990, *Atmos. Environ.* **25A,** 1–203.

Wege, K., Claude, H. and Hartmannsgruber, R., 1989, Several results from the 20 years of ozone observations at Hohenpeissenberg, in: R.D. Bojkov and P. Fabian (eds), *Ozone in the Atmosphere,* Deepak publ. pp. 109–112.

Winiwarter, W., Hawel, R. and Loibl, W., 1994, Spatial and temporal desaggregation of emission inventories as input for air chemistry models - a computer based approach.

EUROTRAC Annual Report part 5: GENEMIS, EUROTRAC ISS, Garmisch-Partenkirchen, pp. 63–67.

Wirtz, K., Roehl, C., Hayman, G.D., Jenkin, M.E., Becker, K.H., Cox, R.A., Le Bras, G., Lesclaux, R., Moortgat, R.G., Sidebottom, H.W. and Zellner, R., 1994, *LACTOZ re-evaluation of the EMEP MSC-W photo-oxidant model.* EUROTRAC ISS, Garmisch-Partenkirchen, 45 pp.

WMO, 1990a, Report of the International Ozone Trends Panel: 1988, *World Meteorological Organization Global Ozone and Monitoring Network Report* **18**, Geneva.

WMO, 1990b, Scientific Assessment of Stratospheric Ozone: 1989, *World Meteorological Organization Global Ozone and Monitoring Network Report* **20**, Geneva.

WMO, 1992a, Scientific Assessment of Ozone Depletion: 1991, *World Meteorological Organization Global Ozone and Monitoring Network Report* **25**, Geneva.

WMO, 1994, Albritton, D.L., R. Watson, and P. Aucamp (eds), *Tropospheric Ozone,* WMO-UNEP Assessment of Stratospheric Ozone Depletion.

Zander, R., Demoulin, Ph., Mahieu, E., 1994a, Monitoring of the atmospheric burdens of CH4, N2O CO, CHCH2 and Cf2Cl2 over central Europe during the last decade, Environmental Monitoring and Assessment, Kluwer Academic Publishers, Dordrecht 31, 203–206.

Zander, R., Ehhalt, D.H., Rinsland, C.P., Schmidt, U., Mahieu, E., Rudolph, J., Demoulin, Ph., Roland, G., Delbouille, L. and Sauval, A.J., 1994, Secular trend and seasonal variability of the column abundance of N_2O above the Jungfraujoch station determined from IR solar spectra. *J. Geophys. Res.,* 99,16475–16456.

Zetzsch, C., and Behnke, W., 1992a, Photocatalysis of tropospheric chemistry by sea spray, *EUROTRAC Annual Report part 6: GCE and HALIPP,* EUROTRAC ISS, Garmisch-Partenkirchen, pp. 121–126.

Zetzsch, C., and Behnke, W., 1992b, Heterogenous Photochemical Sources of Atomic Cl in the Troposphere, *Ber. Bunsenges. Phys. Chem,* **96,** 488–493.

Zimmermann, P.H., 1994, The impact of aircraft-released NO_x on the tropospheric ozone budget: sensitivity studies with a 3D global transport/photochemistry model, in *Proc. Internat. Sci. Colloq. on the Atmosphere,* Cologne, April 1994, DLR-Mitteilung 94-06, 211–216.

Chapter 4

Acidification and Deposition of Nutrients

4.1 Environmental policy issues related to acidic deposition and deposition of nutrients

When developing effect-oriented control strategies for acidification and eutrophication, there are a number of basic questions that have to be answered. These include the severity of the environmental problem, the area affected, its further development, its cause, the methods and costs for controlling emissions and the possibility and time scale of reversibility. Since the overall causes and effects of the acidic deposition are well known, the policy issues today deal mainly with the development and improvement of mechanistic models to analyse the consequences of changes in emissions, as well as the development of cost effective control strategies. The question of whether control measures have been effective also has to be addressed. The scientific research within EUROTRAC has, as this chapter will show, made major contributions in quantifying the processes involved, as well establishing the source-receptor relationships expressed by theoretical models.

In this chapter specific policy-oriented questions, both qualitative and quantitative, will be addressed. The questions are all directed towards the atmospheric processes which include emissions and deposition. Issues related to the impact of pollutants on man, materials and ecosystems, as well as the control techniques and their costs, are not considered since these issues are outside the scope of EUROTRAC.

4.1.1 Establishing links between spatial distribution of sources, deposition and effects

Source-receptor matrices allow deposition in one area to be related to specific sources throughout Europe. Such matrices have played an important role as a basis for the recently signed protocol on sulfur reductions in Europe and are expected to play a similar role for forthcoming international negotiations on control of regional air pollution. Present models provide a reasonable picture of the overall transboundary fluxes and impacts of sulfur and nitrogen in Europe on the country scale. However, more spatial detail in deposition information is necessary especially when considering and establishing source-receptor relationships on national and local scales.

The overall policy question is:

Can source-receptor matrices for acidifying compounds and nutrients be established with sufficient accuracy for the development of effect-oriented cost-effective control strategies?

In addition to the overall question, there are a number of specific questions of crucial importance for the policy and where the knowledge is limited, or at least was limited when EUROTRAC started.

Some of the most important questions are:

Is there an important background in sulfur and nitrogen deposition in Europe caused by natural sources or by sources outside Europe?

The main natural source of acidifying sulfur in the atmosphere is dimethyl sulfide (DMS) emitted from marine ecosystems. DMS may contribute to acidic deposition, at least in coastal areas during summer months. Emissions of nitric oxide (NO) from soils contribute to the background acidity in Europe. Emission estimates require an understanding of the underlying processes in land-atmosphere and ocean-atmosphere exchange to develop suitable models, and such estimates may be important in areas with intensive farming.

Will reductions in emissions of nitrogen and sulfur change the transport scale of these compounds?

In present strategies, it is assumed that emission reductions will reduce sulfur deposition in proportion to the actual source contributions at all sites. Measurements have however identified non-linearities in deposition following the emission reductions in Europe in 1980 (Fricke and Beilke, 1993, MSC-E, 1993). They indicate a faster reduction in the dry deposition of sulfur than expected in central Europe and a slower reduction in wet deposition throughout Europe.

What geographical resolution of deposition is needed to assess environmental effects and exceedances of the critical loads?

Critical loads for sulfur and nitrogen deposition are presently mapped throughout Europe. The exceedance of the critical loads provides a guide to the reduction necessary to achieve sustainable ecosystems. In establishing control strategies it is important to provide estimates of both critical loads and deposition on the same spatial scale which is the appropriate scale to protect the environment.

Does ammonia influence the deposition of sulfur dioxide and vice versa?

Sulfur dioxide is an acidic gas while ammonia is alkaline. During wet conditions, reactions between these gases, either in the atmosphere or at the receptor surfaces, may change the deposition pattern. This co-deposition may influence control strategies. There may be other combinations of reactive gases and particles which influence removal rates at the surface.

Will control of sulfur emissions in Europe significantly influence the radiation balance in the atmosphere?

Climate effects may be local or regional due, for example to the formation of sulfate particles which can act as condensation nuclei, or global due to the cooling effect from sulfate aerosols in the free troposphere. In order to understand the role of European emissions (natural as well anthropogenic), the proportion of these emissions that are transported into the free troposphere needs to be quantified.

Each of the five questions mentioned have, in one way or another, been addressed within EUROTRAC. Some of them, for example the issues of background emissions and the establishment of deposition maps for key species, have been tackled directly, while others have been addressed implicitly. There are of course a number of other questions of relevance for policy that can be raised, such as the role of alkaline emissions for acidification, and the links between control strategies for acidification and photochemical oxidants. These questions have not been the focus of the EUROTRAC project.

4.2   Scientific assessment – issues and highlights

In the following text we describe and discuss the most important scientific developments concerning emissions, atmospheric processes and deposition which are already linked to policy. Special attention is given to the contribution from EUROTRAC. In the presentation, examples of EUROTRAC results, which have been or may be used for the policy process, are provided.

4.2.1   Emissions

The development of source-receptor relationships for acidifying compounds requires emission data with sufficient resolution in time, space and source height, the necessary resolution being determined by the variations in deposition and ecosystem sensitivity that is to be assessed. In Europe, emission inventories were originally made on a grid size of 150 km, the grid size used in the EMEP programme (Tuovinen *et al.*, 1994). Although the grid size is adequate to describe transboundary fluxes in most parts of Europe, it is too coarse to establish links between sources and effects for compounds, where a substantial fraction of the emission is deposited close the source (*e.g.* ammonia), or in complex terrain where large variations in deposition and ecosystem sensitivity are present. Emission inventories within EMEP as well as within the framework of CORINAIR, have been developed with a grid scale of 50×50 km. Nationally, for example in the Netherlands, much finer scales are needed.

Anthropogenic emissions of sulfur dioxide, nitrogen oxides and ammonia from main source categories are, as a part of the LRTAP Convention, subject to yearly national inventories throughout Europe. Before 1987, these inventories were made mostly for sulfur dioxide and their quality has been continuously improved so that the sulfur emission inventories for most of the European countries are considered to be reasonably reliable.

Emission inventories for nitrogen oxides became part of the LRTAP Convention in 1987 and these still contain large uncertainties. Emission inventories of ammonia, although part of the EMEP requirements, are still not made in many countries. Such emission inventories are the work by CORINAIR and EMEP, where the focus is on a more systematic approach; substantial improvements are expected during the forthcoming years.

In addition to the routine emission inventory process, there are important areas where research has improved the inventories, in particular for soil and water bodies, and several of these have been a subject for research in ASE, BIATEX and GENEMIS. In GENEMIS, better time resolution for all sources has been obtained and the knowledge of emissions in the countries of central and eastern Europe has been improved (Friedrich *et al.*, 1994). The effect of the outside temperature on emissions has also been studied. Within ASE and BIATEX, an important task has been to study the emissions of gaseous compounds from different surface categories. These studies include volatile organic sulfur primarily from marine areas (dimethyl sulfide, DMS), while from soils, ammonia from the handling and spreading of manure, and nitric oxide from agricultural soils are the main source terms.

a. Sulfur dioxide

The emissions of sulfur compounds contributing to the deposition in Europe are dominated by anthropogenic sources. The sources are well known, with some exceptions in central and eastern Europe, and most countries in Europe now produce yearly emission inventories.

b. Dimethyl sulfide (DMS)

The oceans make a significant contribution to the global emissions of gaseous sulfur compounds, primarily in the form of DMS. Natural emissions of DMS are thought to play an important role in the production of cloud condensation nuclei in the atmosphere and the emissions may be important in the formation of clouds. It is also believed that global warming may increase the DMS emissions and thus the formation of condensation nuclei resulting in more stable and brighter clouds. These in turn may increase the backscattering of radiation and thus counteract global warming. (Charlson *et al.*, 1987).

It has also been suggested that emissions of DMS may contribute significantly to sulfur deposition and acidification in coastal areas in Europe (Leck, 1989).

The production and consumption of DMS in sea water, and the net emission to the atmosphere have been a focus of several activities within ASE (*e.g.* Belviso *et al.*, 1990; Malin *et al.*, 1993). DMS in sea water is produced from various phytoplankton species: for example, it has been shown that dinoflagellates probably play a key role in determining the abundance of DMS in sea water. These activities have been important links in the understanding of the organisms and the processes regulating emissions. It is clear that anthropogenic sources of sulfur in Europe are responsible for the bulk of the acidifying sulfur deposition while biogenic DMS is important globally.

c. Ammonia

Most of the ammonia emitted from anthropogenic sources originates from livestock manure and its handling (including spreading). Other but less important sources are the manufacture of ammonia and nitrogen fertilisers, the application of chemical fertilisers and from SCR flue gas treatment. The emissions of ammonia, although having similar ecosystem effects when deposited as oxidised nitrogen, have received much less attention in terms of quantification. Emission inventories for ammonia are necessary for the assessment of acidification and eutrophication effects and the implementation of the critical load concept. A first estimate of the surface emissions in Europe was made in 1982 (Buijsman, 1987). This emission

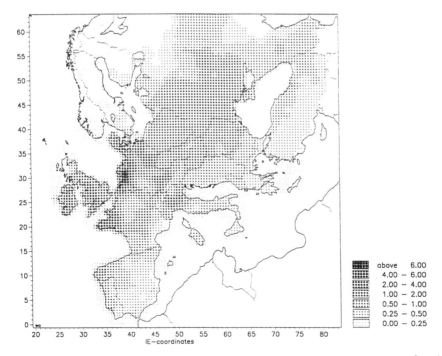

Fig. 4.1:   Emissions of ammonia in Europe 1989 expressed as tonne NH_3 km^{-2} yr^{-1} (Asman, 1992).

inventory from joint work between ASE and GENEMIS (Fig. 4.1). has been updated (Asman, 1992), and it clearly identifies the role of farming as the source of nitrogen-related environmental problems in parts of Europe.

The total annual emission of nitrogen from ammonia, estimated to be 7.6 million tonnes for 1990 by Asman, is comparable with the annual emission of nitrogen from NO_x of 7.1 million tonnes, as reported to the LRTAP Convention (Tuovinen et al., 1994). Although the emission inventories still lack accuracy, it is obvious that the emissions of reduced nitrogen are as important as those of oxidised nitrogen for the biosphere. The differing roles of NO_x and NH_3 are considered in later sections of this chapter.

d. Nitrogen oxides

The emissions of NO_x (NO and NO_2) are dominated by anthropogenic sources, of which traffic and stationary combustion are the main contributors. The sources are well known, although the emission inventories still lack accuracy. In addition to these major sources, it has been suggested that soils also contribute on the continental scale.

The emission of nitric oxide from soils was investigated by laboratory and field measurements as a part of BIATEX. The results show that agricultural soils generally emit appreciable amounts of nitric oxide (Remde and Conrad, 1991). The results obtained have however not yet provided detailed figures for Europe as a whole, but initial estimates show that soil emissions are an important additional source of NO_x in agricultural areas (Skiba et al., 1992; 1994; Remde and Conrad, 1991). The emissions vary substantially with climate and agricultural practices, and process studies provide the basis for estimating annual emissions over large areas of Europe.

4.2.2   Atmospheric processes

a. Introduction

Most gases entering the atmosphere from natural and anthropogenic sources are in a chemically reduced state and undergo oxidation in the atmosphere. Typical examples are SO_2 and NO emitted from combustion sources. Sulfur dioxide in the atmosphere is oxidised to sulfuric acid, whereas NO is oxidised first to NO_2 and thereafter to HNO_3. These chemical processes compete with removal at the ground for the primary pollutants. Chemical conversion causes a change in the properties of the substance, which leads to changes in deposition rates, life-times and transport distances. The combination of chemical conversion and dry deposition therefore determines the transport distance between sources and their effects. For example, whereas the dry deposition of NO is an inefficient process, dry deposition of HNO_3 is extremely rapid (Müller et al. 1993). Another important

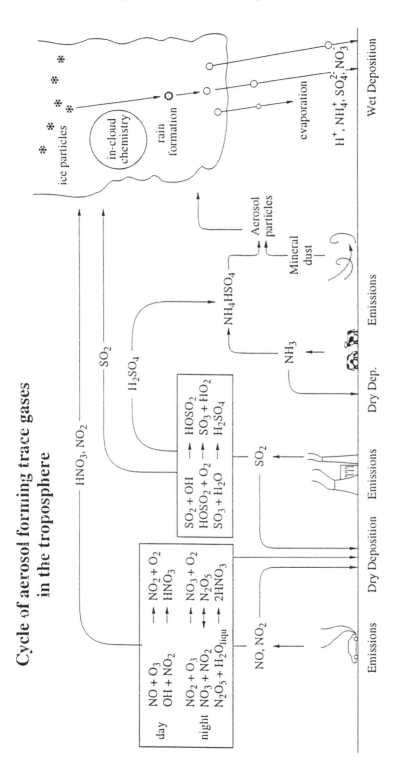

Fig. 4.2: The connections between gas-phase processes, aerosols and clouds in the formation of nitric and sulfuric acid (after Peter Warneck).

aspect is that nitric acid and sulfuric acid, which are the final oxidation products of gaseous nitrogen and sulfur compounds in the atmosphere, have a tendency to associate with alkaline elements or compounds and form salts. These processes contribute to the formation of particulate matter in the atmosphere. The role of ammonia in the neutralisation of sulfuric acid and the formation of ammonium sulfates and nitrates is well recognised. Aerosol particles assist in the generation of clouds in that they serve as cloud condensation nuclei (Hobbs, 1993). Contributions to wet acid deposition come from particulate matter incorporated in clouds as well as from materials formed in clouds as a result of chemical reactions. Fig. 4.2 illustrates the connections between gas-phase processes, aerosols and clouds in the formation and wet deposition of nitric and sulfuric acid.

At the beginning of EUROTRAC, regional and global transport models incorporating gas-phase chemical reactions were available and widely utilised. Examples are EMEP and PHOXA, which are described elsewhere in this report (see Section 4.4). These models describe mainly oxidant formation and deposition processes. The EURAD model applied in EUROTRAC uses a highly parameterised cloud chemistry mechanism adopted from RADM (Chang *et al.*, 1987, Hass, 1991). It was however clear that the scientific knowledge about the uptake of gases by cloud droplets and of the detailed reactions within the droplets was insufficient. The subproject HALIPP specifically addressed this problem and has, in addition, provided a better understanding of nucleation processes involved in cloud formation, while the subproject GCE set out to study physical and chemical processes occurring in clouds, in the field. Much of the laboratory work has emphasised chemical mechanisms of sulfur dioxide oxidation in clouds. The progress achieved in this field of research will be summarised below. Ultimately, a quantitative description of in-cloud oxidation of sulfur dioxide will have to be included in transport models in order to solve the problem of the transport distance of sulfur dioxide.

b. Gas-phase chemistry

Gas-phase reactions have been studied extensively during the last two decades and many of the conversion processes of interest were reasonably well defined at the outset of EUROTRAC (Atkinson *et al.*, 1989, Calvert and Stockwell, 1984, Warneck, 1988). An exception is the oxidation of DMS for which details of the oxidation mechanism were, and still are, rather uncertain (Restelli and Angeletti, 1993). As the ocean is the main source of DMS on a global scale and because DMS in the atmosphere has a short chemical life-time, the formation of acidic sulfur compounds from DMS will be important primarily in the marine atmosphere.

Reactions involved in the gas-phase conversion of nitrogen oxides and sulfur dioxide that usually are included in transport models are shown schematically in Fig. 4.2. Both oxides are known to react with OH radicals. In the case of nitrogen dioxide the reaction with OH leads to nitric acid directly. Sulfur dioxide, in reacting with OH, first forms an adduct, which interacts with oxygen and water to produce sulfuric acid (Stockwell and Calvert, 1983). Nitrogen dioxide also is converted to nitric acid at night. This route involves the reaction between nitrogen dioxide and ozone forming nitrogen trioxide, followed by the addition of nitrogen dioxide to generate nitrogen pentoxide, which ultimately reacts with liquid water to produce nitric acid. Both reaction pathways are almost equally effective, leading to a life-time of nitrogen dioxide in the atmosphere of 1 to 2 days. The life time of sulfur dioxide based solely on its reaction with OH radicals is longer and is estimated to be about 12 days. The actual life time of SO_2 is shorter, of the order of one day, due to other removal processes such as dry deposition and oxidation in clouds (Warneck, 1988).

c. Atmospheric chemistry in clouds

The following summary is subdivided to highlight results from laboratory studies and field observations. The development of cloud models will be discussed further below.

The modelling of clouds is complicated by the presence of at least one additional phase, that of water, and the exchange of materials between phases. Water soluble materials are partly transferred to the liquid phase. The time constant for transfer depends on drop size, the solubility of the substance involved, and the phase transfer probability commonly expressed as a mass accommodation coefficient. Within HALIPP mass accommodation coefficients have been measured for a number of key species specified later (Bongartz and Schurath, 1993; Ponche et al., 1993; Bongartz et al., 1994; Bongartz et al., 1995) making it possible to calculate transfer rates for these species.

Clouds form by condensation of water onto aerosol particles that serve as condensation nuclei. Although the nature of the process was well understood when EUROTRAC started, the efficiency of heterogeneous nucleation depends partly on the chemical composition of the particles, and it was not known whether the assumption, made in models, of an internal chemical mixture represented the actual situation sufficiently well.

Gas-phase photochemistry induced by sunlight continues in clouds in a similar manner to cloud-free conditions. However, as strongly water-soluble components produced by photochemical reactions, such as HO_2 radicals and the combination product H_2O_2, are transferred from the gas to the aqueous phase, the extent of gas-phase chemistry changes. One of the consequences appears to be a reduction in the rate of ozone formation (Lelieveld and Crutzen, 1991). At the same time the species entering the cloud water undergo reactions in the aqueous phase. A well-recognised problem is the oxidation of anthropogenic sulfur dioxide in the aqueous phase, where it exists mainly in the form of hydrogen sulfite and sulfite anions, often denoted as S(IV). It has been estimated that clouds contribute from 48 % up to 84 % to the conversion of SO_2 to H_2SO_4 in the troposphere (for a comparison see Langner and Rodhe, 1991). Appreciable progress has now been made in identifying the reactions responsible for S(IV) oxidation in the aqueous phase and in determining the associated rate coefficients. As indicated in Fig. 4.3, four processes cause the oxidation of S(IV) species in the aqueous phase of clouds: reactions with peroxides, with ozone, and with oxygen when catalysed by OH radicals or transition metals (Warneck, 1988, 1991). The reaction with hydrogen peroxide usually is considered the leading reaction.

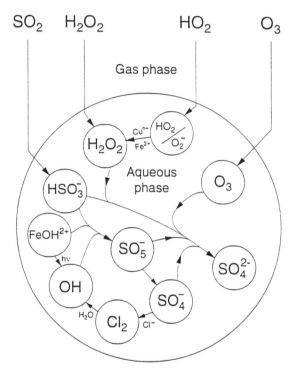

Fig. 4.3:   The processes causing the oxidation of S(IV) species in the aqueous phase of clouds. Briefly, these are reactions with peroxides, with ozone, and with oxygen when catalysed by OH radicals or transition metals.

There is also evidence that surface reactions on aerosol particles are important in the conversion of certain gaseous species in the atmosphere. These processes needed closer scrutiny by appropriate laboratory investigations.

The purpose of field observations of clouds is to test concepts developed to describe the microphysical and chemical processes associated with clouds. Although field measurements had been performed earlier, they were generally limited in scope. GCE offered for the first time an opportunity to conduct integrated measurement campaigns involving larger research groups, each handling specialised equipment, so as to obtain more complete data sets by a joint effort. Limitations also existed with regard to measurement techniques. Groups active within GCE made a significant effort to improve or develop new experimental methods. The primary targets for study were the process of nucleation and the contribution of particulate matter to the chemical components occurring in cloud drops; the dissolution of gases in cloud water and the adjustment to Henry's law equilibrium; the oxidation of sulfur(IV) in the aqueous phase of clouds; and losses by wet deposition of cloud or fog drops to the ground (Heintzenberg, 1992; Fuzzi, 1994; Pahl *et al.*, 1994).

d. Results from laboratory studies

The studies of wet chemistry processes within EUROTRAC have mainly been directed towards the importance of cloud processes. The main scientific highlights from the research are listed below.

Mass accommodation coefficients

- Gas-to-aqueous phase mass accommodation coefficients have been determined for important solutes such as HNO_2, HNO_3, N_2O_5, NH_3, HCl, SO_2, HCOOH, O_3; most values are of the order of 0.01.

Reactions oxidising SO_2

- The direct oxidation of HSO_3^- and SO_3^{2-} by H_2O_2 is a non-radical process. Mechanism and rate coefficient as a function of pH have been fully established for this important reaction.

- The reaction of S(IV) with ozone has been better defined under conditions of high ionic strength relevant to deliquescent cloud droplets. However, the important question whether the reaction involves radicals or not has not been answered.

- The mechanism of the OH radical induced chain oxidation of S(IV) has been fully established, and rate coefficients were obtained for the chain initiating, chain propagating and chain terminating reaction steps. Chain carriers are SO_5^- and SO_4^- radicals. Another important intermediate is peroxomonosulfate. Its reaction with HSO_3^-/SO_3^{2-} was studied and the rate coefficient determined as a function of pH.

- S(IV) oxidation catalysed by ionic iron, which has been under investigation for almost a century, has been shown to proceed by a chain reaction involving, as one elemental part, the propagation and termination reactions active as well in the OH induced oxidation reaction. OH radicals are not generated. The rate coefficient for the initiation reaction between Fe(III) and S(IV) has been defined, and other important reactions with Fe(II) and Fe(III) species have been identified.

- Effects of other catalytically active transition metals such as manganese have been studied and interpreted in terms of a mechanism similar to that of iron. Verification will require more effort, however.

Sources of OH radicals

- The source of OH radicals in cloud droplets has not been easy to identify. The entrainment of HO_2 radicals, dissociation to form O_2^-, and its reaction with ozone, which was thought to provide a mode of OH production, probably cannot compete with the scavenging of HO_2/O_2^- by transition metals. Copper ions appears to be the most efficient scavengers, converting HO_2/O_2^- either to H_2O_2 or O_2 depending on the oxidation state of copper.

- Among the aqueous substances investigated that may produce OH by photodecomposition (NO_3^-, HNO_2/NO_2^-, H_2O_2, $FeOH^{2+}$, $FeSO_4^+$), the Fe(III)-hydroxo complex is most efficient in terms of quantum yield and abundance in cloud water.

Heterogeneous reactions

- Rates and products of N_2O_5 reacting with deliquescent and dry salt particles have been determined. This is an example of an aerosol surface reaction of importance to the conversion of NO_2 to HNO_3 in the atmosphere.

The knowledge of aqueous-phase chemistry relevant to clouds has advanced to point where it is possible to construct reaction mechanisms for application to clouds in atmospheric simulation models. This research activity should now be given priority, even though efforts to reduce uncertainties in reaction mechanisms should continue, especially with regard to the role of transition metals (for details, see Volume 2 of this series).

e. Results from field observations

A variety of clouds occur in the atmosphere, with convective and stratiform types being the most frequent. Ground-based field measurements require not only suitable observation sites (usually hilltop locations) for the study of clouds, but additionally an infrastructure to support the measurements. Three measurement sites were utilised to study different types of cloud systems:

(1) Fog in the Po Valley, Italy, which was characterised by extremely low updraft velocity, high pollutant concentrations, and an oxidant-limited situation;

(2) Stratus clouds encountered at the mountain station of Kleiner Feldberg, Germany, which were characterised by moderate updraft velocity, high pollutant and high oxidant concentrations;

(3) Orographic clouds encountered at Great Dun Fell, UK, which were characterised by high updraft velocity, low pollutant and high oxidant concentrations.

The major results and conclusions available to date are as follows:

- Nucleation scavenging is the principal mechanism for the incorporation of particulate matter in cloud drops;

- Two groups of particles having different hygroscopic and therefore different growth properties were identified in all experiments (at each location). The more hygroscopic particles are scavenged to a greater extent than the less hygroscopic ones. Populations of both types of particles exist independent of size. The hygroscopically inactive particles may consist largely of organic material;

- Scavenged and interstitial aerosol particles are significantly different in chemical composition. The efficiency of scavenging of elemental carbon, for example, is 30 % smaller than that of sulfate;

- Chemical composition and solute concentration in cloud drops is size-dependent. Concentrations may either increase or decrease with increasing drop size. Both cases have been observed, the reasons for the differences being found in the different saturation histories of the respective air parcels;

- Discrepancies of up to two orders of magnitude have been observed between Henry's law predicted and measured gas-liquid-phase partitioning of moderately soluble gases such as ammonia and organic acids. This effect may be due to a rather inhomogeneous distribution of pH among individual droplets and the bulk water sampling procedures mostly employed;

- The acidity of cloud and fog water is mainly controlled by the availability of gaseous ammonia within the cloud system. Peak acidities are connected with the input of nitric acid to the system. It is not yet clear whether nitric acid is produced within the cloud or admixed with air from source regions;

- Under the oxidant-limited conditions of the field measurement campaigns in the Po Valley and at Kleiner Feldberg no appreciable conversion of sulfur dioxide to sulfate was observed. However, sulfate production has been observed at Great Dun Fell.

Field measurements are important in two ways: on the one hand they are necessary to provide support for existing concepts and/or models of the microphysical and chemical processes involved in cloud growth and behaviour, on the other hand they can identify areas where existing knowledge is inadequate and better concepts must be developed. One important observation made in recent years is a considerable inhomogeneity in chemical composition among cloud drops, which appears to result from the chemical diversity of nucleating particles. This effect has not yet been taken into account in cloud models.

4.2.3   Non-linearities in source-receptor relationships

The development of abatement strategies in Europe are based on the assumption that changes in emissions will result in similar changes in deposition, taking into account the relative importance of the emissions concerned. However observations only partly support this assumption. Measurements in Germany on concentrations of sulfur compounds in the atmosphere and sulfur deposition by precipitation indicate that the emission decreases in western Europe were accompanied by a much larger decrease in the concentrations of sulfur dioxide, compared to the decreases in concentrations of sulfate aerosol and sulfur deposition by precipitation (Fricke and Beilke, 1993). Similar conclusions were drawn from data from a number of EMEP stations (MSC-E,1993). These evaluations indicate that emission reductions achieved in western Europe between 1980 and 1990 gave a much larger reduction in the concentrations of sulfur dioxide (and thus the dry deposition of SO_2) compared to the reduction in wet deposition. The downward trend in the SO_2 concentration was of the order of 3 to 5 % per year, while the overall downward trend in wet deposition was about 0.5 % per year. During the same time the overall downward trend in emissions was about 2.5 % per year. Such large differences between trends in wet and dry deposition shown by EMEP data may have a substantial influence on control strategies.

The large differences observed in trends between wet and dry deposition together with some of the results from EUROTRAC highlight the issues of non-linearity and may therefore make an important contribution to the assessment of the importance of these processes and also provide a basis for including these processes in future models. Cloud processes seem in this context to be very important.

Cloud processes, in contrast to most gas-phase chemical processes, tend to be highly non-linear. The origin of the non-linearity lies in the nature of the process of cloud formation, but cloud chemistry may also be a factor. Laboratory data show that the oxidation of sulfur (IV) species in aqueous solution in the presence of transition metals can be autocatalytic, that is the rate of the reaction increases with the accumulation of the products (Bal Reddy and van Eldik, 1992; Berglund *et al.*, 1993; Pasiuk-Bronikowska and Rudzinski, 1993; Ziajka *et al.*, 1994). The main cause of autocatalysis is a branching chain reaction but it is not yet known whether this feature is retained under conditions existing in clouds or fogs. Another possible cause is the direct oxidation of transition metals by oxygen (Pasiuk-Bronikowska and Rudzinski, 1993b, 1994). Gas-phase chain reactions leading to ozone formation during the oxidation of hydrocarbons in the atmosphere, occur in a concentration regime where the process is linear. It is possible that this also happens with chain reactions in cloud droplets.

Non-linearities also appear when the supply of oxidant is limited. In this case, the oxidant is fully consumed when the reactant to be oxidised is present in large excess. For example, H_2O_2 produced photochemically in the gas phase and entering the liquid phase is expected to be completely consumed in oxidising dissolved SO_2, when the concentration of the latter is high. If this were the major reaction for in-cloud SO_2 oxidation, the production of H_2O_2 would be rate limiting. Because parameterisation schemes used in most models previously linearised the SO_2 oxidation process, these models could not cope with the problem of non-linearity. A recent GLOMAC study based on a global transport model involved the opposite assumption, namely that SO_2 is completely converted to sulfate whenever air containing sulfur dioxide is cycled through a cloud (Langner *et al.*, 1994). The results are broadly consistent with observations, but the assumption of complete SO_2 conversion is rather unrealistic. In this study the concentrations of sulfur dioxide were assumed to be rather low as is typical for the marine environment. Over the continents the concentrations are larger, and here the concentrations of oxidants are expected to be rate limiting. The true rates, the degree of conversion of SO_2 into sulfate in clouds, as well as the occurrence of non-linearity in this process, remain a problem to be resolved after the cloud chemistry is satisfactorily included in transport models.

Grennfelt *et al.* (1994) have pointed out that a reduction in precursor emissions of a particular pollutant may influence the transport distance of other pollutants. For example, when NO_x or organic emissions are reduced, one expects lower rates of formation of oxidants such as ozone. This in turn would lead to reduced OH radical formation rates causing a decrease in the rate of oxidation of sulfur dioxide in the gas phase. A drop in the formation rate of H_2O_2 may also decrease the rate of sulfur dioxide oxidation in the aqueous phase of clouds. If the rate of SO_2 oxidation in clouds were limited by the availability of oxidants such as hydrogen peroxide, less sulfur dioxide would undergo oxidation and this would increase the SO_2 transport distance. Another effect is that caused by the interactions between ammonia and sulfur dioxide in the atmosphere. Gaseous ammonia will be more easily deposited than ammonium ions in aerosols (sulfates and nitrate). Reductions in ammonia emissions in areas with high ammonia emissions will therefore probably give larger reductions in deposition in the source areas than at greater distances.

4.2.4   Deposition

Until the end of the 1980s, the deposition rates for many of the major pollutants were poorly known and were based on studies over very few land types. Sulfur dioxide was the only gas for which a substantial quantity of data were available. The application of the understanding available at that time was limited largely because of the lack of a detailed mechanistic model to estimate area inputs. Applications were restricted to use of a single deposition velocity to represent inputs on regional scales. For other gases such as NO_2 and for particles, there were no field measurements of deposition rates and the values of the deposition velocity used in long range transport models were very uncertain. Instead, deposition estimates for NO_2 and particles were primarily based on laboratory studies.

Wet deposition was at that time well monitored throughout large areas of northern Europe, particularly in the regions experiencing the main ecological effects of deposited acidity. There were however areas in which wet deposition was also very uncertain, either because of complexity in the terrain, sampling or interpolation difficulties or because the input process was not detected by the monitoring equipment, as in the case of cloud water deposition.

The range of compounds and processes covered by EUROTRAC includes the major gaseous sulfur and nitrogen compounds, the major processes regulating surface-atmosphere exchange and a focus on terrestrial land types, which are representative of the natural and managed ecosystems of Europe. The focus of the measurement activity was directed toward the compounds for which little or no field data were available or about which there were major uncertainties. These included NO, NO_2, NH_3, HNO_3 and related species closely connected with the formation and occurrence of photochemical oxidants O_3, H_2O_2 etc. The deposition of aerosols and cloud droplets were included but were a secondary aspect of the field activities. The measurement programme also included SO_2, but in this case the primary reasons for its inclusion were to investigate the interaction between SO_2 and NH_3 within the deposition process and to validate models by examining the variability in key variables (notably canopy resistance) under a range of conditions.

The following paragraphs highlight specific aspects of the EUROTRAC deposition research which are important for policy development (Slanina et al., 1994). Most of the activities were undertaken within BIATEX and ASE.

a. Deposition of NO and NO₂ to terrestrial ecosystems

Prior to EUROTRAC, studies of rates of NO_2 deposition under field conditions were scarce and the results contradictory. During the period between 1985 and 92, a number of investigations were directed towards the dry deposition of nitrogen species, several of them as part of BIATEX. The BIATEX programme included large field campaigns to cover all the chemical species and key mechanisms. Using this approach, the NO_2 deposition process was shown to be regulated by the stomatal resistance of vegetation, leaf and soil surface uptake rates being negligible under most conditions (Hargreaves et al., 1992; Rondon et al., 1993; Granat 1993). The dominating role of the stomatal pathway simplifies parameterisation of the deposition and makes it possible to include a mechanistic approach for the quantification of NO_2 deposition in various models.

A major problem in quantifying the deposition of NO and NO_2 has been the modification of the concentration profiles due to chemical reactions, primarily with ozone. This modification has been a continuing problem in this field for some time. Within BIATEX important improvements have been made in the treatment of these effects (Duyzer et al., 1994; Kramm et al., 1991) and, as a result of this progress, the real fluxes have been separated from artefacts of the experimental methods.

The field data on NO_2 deposition provide a tool for making quantitative estimates of the dry deposition of NO_2 to terrestrial ecosystems. The data have been used within resistance models of the deposition process, with NO_2 monitoring, land use and climatological data to provide fine scale dry deposition estimates within the UK and the Netherlands (Erisman *et al.*, 1993a; Duyzer and Fowler, 1993; Coe and Gallagher, 1992; Meixner and Ludwig, 1992). These developments are now available for implementation on the European scale.

b. Surface-atmosphere exchange of NH₃

The surface-atmosphere exchange of NH_3 is complicated by bi-directional exchange processes (emission and deposition) which result from the interaction between NH_3 in the atmosphere and the biological formation and processing of NH_3 within plants and soils. The programme of field measurements of NH_3 fluxes over natural and managed land and over forests using micrometeorological methods has provided an understanding of the main factors which regulate the process (Duyzer *et al.*, 1987; Wyers *et al.*, 1993; Sutton *et al.*, 1995). These developments have led to the first approaches to modelling the net exchange of NH_3 fluxes over vegetation (Sutton *et al.*, 1993). The inclusion of the deposition/emission relationship tends to smooth out the deposition of ammonia in time and space. Direct measurements of ammonia exchange have been made in an ASE project on eutrophication in Danish waters.

c. Interactions between nitrogen and sulfur compounds in the deposition process at natural surfaces (NH₃ and SO₂ co-deposition)

Field measurements have provided evidence of interactions between SO_2 and NH_3 on leaf surfaces which is consistent with the concept of co-deposition. The process leads to enhanced deposition of SO_2 in the presence of NH_3 and vice versa. The process has been shown to be more complex than earlier laboratory studies had implied (Sutton *et al.*, 1991) and the precise details of the mechanisms have yet to be elucidated. The consequence is that policies to control either of the reactants will influence the deposition and the 'effects footprint' of the other. Integrated field experiments and monitoring of dry deposition fluxes over annual time scales have been undertaken in the Netherlands (Wyers *et al.*, 1993).

d. Provision, application and validation of mechanistic models to estimate the dry deposition of SO₂ to a wide range of terrestrial surfaces

The understanding of deposition processes has been incorporated in process-based mechanistic models to provide inputs on field, catchment and national scales. This activity has taken place throughout the period of EUROTRAC in parallel with the field studies (Erisman, 1994, Erisman *et al.*, 1990; 1993a; 1993b; Erisman and Wyers, 1993). The development of methods for continuous monitoring of deposition fluxes has provided the means of validating long-term annual deposition estimates (Erisman and Baldocchi, 1993, Erisman *et al.*, 1993a). Catchment budget methods have also provided a valuable technique for validating long-term deposition estimates (Reynolds *et al.*, 1994). Comparisons with receptor-oriented monitoring methods (primarily throughfall and catchment studies) show, however, that present deposition calculations are only partly able to explain the variability in ecosystem fluxes. (Draaijers and Erisman, 1993; Erisman *et al.*, 1993a)

e. Deposition of aerosols

Research on the deposition velocities of aerosols was added to BIATEX at a late stage. The reason for doing so was the large differences in deposition velocities derived from theoretical calculations and wind tunnel experiments compared with the few values obtained by field experiments (Ruigrok *et al.* 1993). While theory and wind tunnel experiments indicated low deposition velocities, of the order of 0.1 cm s^{-1} or less for sub-micron particles, the few results of field studies led to estimates of 0.5 cm s^{-1} over low vegetation and values of 1 cm s^{-1} or higher over forests. Improved methodology and the application of tracers has been applied in field experiments carried out over the last three years. The results of different research groups (Lopez, 1994) agree within an acceptable range of uncertainty and lead to the conclusion that the high values reported by early field experiments are correct. Deposition velocities of sulfate and nitrate particles are in the order of 0.5 cm s^{-1} over low vegetation and deposition velocities of over 1 cm s^{-1} are found for forests. These findings are important in describing inputs to ecosystems and long range transport of pollutants. The life-time of aerosols has become a major issue in view of the strong influence of aerosols on the radiative budget of the earth, and the loss by deposition of aerosols is a major factor in understanding the radiative influence of small particles.

The related problem of the deposition velocities over water was the subject of an ASE experimental study of the influence of sea spray and white caps on the deposition velocity. The effects were found to be small and the classical expressions can be applied (Slinn, 1992). Thus, unlike the BIATEX results for low vegetation, wind tunnel results can be applied to natural water surfaces (Larsen *et al.*, 1993).

f. Deposition of acidifying substances by cloud water deposition

It has been demonstrated that deposition by cloud water may make a significant contribution to the total deposition of acidifying compounds, especially at sites with a high frequency of clouds and fog. This is especially the case at elevated sites which are often above the cloud base. The deposition pathway has been a focus for scientific research throughout the last decade. The research includes identifying the mechanisms which regulate cloud water deposition and the effects of orography on the wet deposition process, developing models and using the models to provide national maps.

Simulation of long range transport, carried out within EUMAC, indicate that a better understanding of ammonium sulfate and ammonium nitrate aerosol formats is indispensable for explaining acid deposition in remote areas (Ziegenbein *et al.*, 1994).

The BIATEX programme included a series of field studies of the interactions between aerosol, fog and cloud-droplet composition and the distribution of the key chemical components between the different phases. The work also included studies of the deposition of cloud droplets and the contribution of the process to acidification of ecosystems. In particular it demonstrated the size dependence of the deposition process and has provided the basis for a model to estimate inputs of sulfur and nitrogen on a regional scale (Dore *et al.*, 1992). While this process makes only a minor contribution to annual budgets, it has been shown to be important in some of the most sensitive ecosystems, the high elevation forests, and has important implications for afforestation programmes in acid sensitive areas.

g. Quantifying the atmospheric inputs of nitrogen to coastal waters

The dry deposition of nitrogen compounds to sea surfaces includes some unusual processes which must be taken into account when developing models for dry deposition. One of these is the association of nitric acid with sea salt particles, which may change the deposition properties radically (Ottley and Harrison, 1992). Within ASE, dry deposition models for nitrogen species have been developed (Asman *et al.*, 1994a). The measurement and modelling of wet and dry deposition of nitrogen compounds to the coastal waters of the Kattegat has provided the first detailed assessment of the scale of these inputs and the magnitude of the individual components (Asman *et al.*, 1994b). Although the wet deposition inputs are dominant, there is an important contribution to the total from dry deposition of NH_3. The dependence of nitrogen deposition on meteorological conditions has been demonstrated with the EURAD model (Ebel *et al.*, 1993; see section 5.4.2).

h. Identifying interactions between the deposition of sulfur and nitrogen compounds and emissions of the radiatively active gases CH_4 and N_2O

The large field studies of fluxes of various nitrogen trace gases included N_2O which is one of the gaseous products from the microbial processing of available nitrogen. The emissions measured represent a small but appreciable fraction of the gaseous emissions (Skiba *et al.*, 1992) and are illustrative of the links between the deposition fields for nitrogen compounds and emissions of radiatively active gases. There is also new evidence of effects of deposited sulfur on the emission of CH_4 from wetlands (Fowler *et al.*, 1994a; Hargreaves *et al.*, 1994). The data indicate a decreased CH_4 emission from peat wetlands following deposition of SO_4^{2-}.

The activities within BIATEX and ASE have contributed substantially to the overall understanding of deposition mechanisms and their parameterisation for use in modelling the behaviour of sulfur and nitrogen compounds. The achievements have been most significant for gaseous sulfur and nitrogen compounds (Erisman *et al.*, 1994). Much of the research has been directed towards an integrated approach, where atmospheric as well as internal receptor processes are taken into account.

4.3   Applications and importance for policy

The goal of the Application Project has been to identify the contribution from EUROTRAC during the last decade to the development of air pollution policies in Europe. Scientific research on air pollution effects and the baseline monitoring of the air pollutants are largely outside the remit of EUROTRAC, but the links to these areas have been important for the success of some of the EUROTRAC projects, for example BIATEX, TOR and ASE, which have provided information for the assessment of effects.

To understand the processes involved in the atmosphere and to develop policies, mechanistic models must be applied. Practically all mechanistic and process-oriented research on the deposition of acidifying compounds and nutrients within EUROTRAC has been done with the objective of supporting models either in their parameterisation or in their validation. It should however be noted that such models are simply the tools for generalisation and integration of the best available knowledge. They are not a substitute for monitoring and field experiments directed towards understanding processes and quantifying amounts of pollutants transported over Europe. Since the development of parameters for use in models have been such an important part, the review of applications will focus on this point.

Atmospheric models, which quantify the links between sources and receptors, include descriptions of the emissions, the atmospheric processes (chemistry, transport and dispersion) and the deposition processes. In our evaluation we will review each of these processes and identify areas where knowledge has increased during the last 10 years and where EUROTRAC has contributed substantially.

4.3.1   Quantification of emissions

a. Dimethyl sulfide

Emission estimates and model calculations indicate that DMS emissions from marine areas such as the North Sea and North Atlantic can make up 5 % of the yearly anthropogenic emissions within Europe (Tarrason, 1991) (Table 4.1). Deposition estimates show that DMS emissions may give an appreciable contribution to the overall deposition in the most western parts of the British Isles, approaching 10 % of the yearly deposition (Touvinen et al., 1994). For Scandinavia the contribution is much less and during summer months, when the contribution from DMS is highest, it may give a maximum contribution to the deposition in the order of 10 % (Leck, 1989). On a yearly basis, the contribution from DMS to the deposition in Scandinavia is of the order of 1 %. The figures on emissions and the role of oxidation and deposition have not yet been fully evaluated and the estimates in Table 4.1 must therefore be considered as provisional.

Table 4.1:   Estimates of global and European, annual gaseous sulfur emissions (IPCC 1994; Touvinen et al., 1994; Tarrason, 1991). In the European emissions, DMS emissions from the North Sea and North Atlantic are included.

Source	Global emissions Tg S yr^{-1}	European emissions Tg S yr^{-1}
Anthropogenic (mainly form fossil fuel combustion) (SO_2)	80	20
Biomass burning (SO_2)	7	< 1
Marine biogenic emissions (DMS)	40	1
Soils and plants (H_2S)	10	< 0.1
Volcanoes (H_2S, SO_2)	10	< 1
Total	147	21

b. Ammonia

Within the LRTAP Convention, all countries are obliged to report their yearly emissions of ammonia. Approximately half the countries under the Convention had in the spring of 1994 still not made their own emission inventories. Instead, the data from other estimates had to be used (Touvinen et al., 1994). The emission inventories compiled by Asman (1992), as a part of GENEMIS, have been an important source in estimating missing emissions as well as for validating nationally calculated emissions (Asman, 1992). Data from this inventory have been used in other models, and to determine national policies. Data from the emission inventory, in combination with monitored and modelled data on distribution and effects, were the main basis for the Dutch policy on reduction of ammonia emissions currently being implemented. The main policy implication of these emission inventories is that ammonia needs to be considered as important as the nitrogen oxides when combating acidification and eutrophication.

c. Nitrogen oxides

The scattered results on emissions of nitrogen oxides indicate that emissions from fertilised land are of the order of a few kilograms of N per hectare and year. If however present data are integrated to a European scale, estimates indicate that the emissions are of the order of 5 % of the total European NO_x emissions and would become important for policy regarding NO_x in Europe should emissions from industry and transport decline (Williams et al., 1992). Laboratory work (Remde and Conrad, 1991) has indicated the main controls on rates of NO production and emission for a range of soils. However a process-based model appropriate to estimate regional emissions has not yet been developed. A few attempts to scale NO emissions are limited to empirical models based largely on soil NO_3^- and temperature (Skiba *et al.* 1994) and applied at country scale by Williams *et al.* (1992).

4.3.2   Cloud chemistry

As mentioned earlier, cloud processes are thought to contribute at least 50 % of the atmospheric oxidation of sulfur dioxide. Quantitative modelling of the cloud oxidation processes are therefore of particular importance for the overall quantification of source-receptor relationships. Thus, the rationale for laboratory and field experiments of chemical processes in EUROTRAC was to obtain information for inclusion in chemical transport models simulating the fate of pollutants in the atmosphere over Europe. Most transport models include an appropriate parameterisation scheme for wet deposition as a loss process for trace species. For this purpose, it is not necessary to treat cloud processes in detail. A lumped chemical conversion parameter for the oxidation of SO_2 and NO_2 can also be included in the parameterisation. This type of treatment, however, precludes the implementation of more complex chemical reaction sequences similar to those used to describe gas-phase chemistry, and it may lead to false conclusions. One of the aims of EUROTRAC was to provide a detailed list of reactions occurring in the aqueous phase. The results have now opened the road to a more realistic description of in-cloud chemistry, specifically the oxidation of SO_2 in clouds deemed to be so important. It remains to include this new knowledge into transport models, where the parameterisation schemes used up to now will have to be replaced by computational modules better suited to the problem.

In recent years, more realistic cloud models have become available that treat cloud-droplet evolution, growth and coalescence, and follow the distribution of trace species within the drop size spectrum and in precipitation. The models are frequently so demanding of computer resources that they do not allow complex chemical reaction sequences to be included. On the other hand, there are models designed to treat chemical reactions in detail, but they work with a standard uniform drop size, which is kept stationary in time. Intermediate type models also have been developed. These models have been applied primarily in research, not for policy purposes. Floßmann *et al.* (1994a) have produced a very useful compilation of cloud models recently developed in Europe, but many of the models were designed to obtain a better understanding of cloud microphysics rather than chemistry. Table 4.2 presents a summary description of several cloud models that incorporate at least some chemistry. The most advanced of these is the FOG model by Bott and Carmichael (1993), which includes the full range of thermal fluxes, microphysics and chemistry. The chemical mechanism probably requires updating. In addition it will be clear that radiation fogs, for which the model was developed, cannot substitute for other cloud types that occur more

Table 4.2:   Cloud models including chemistry (extracted from Floßmann, 1994a)

Model Acronym Cloud Type	Purpose of model (Ref.)	Physical Description	Chemical Description
EURAD Convective cloud	Simulate transport, chemical transformation and deposition of pollutants (Hass, 1991, Mölders et al., 1994)	3-D Eulerian 3000×3000 km zooming to smaller scale; cloud water accumulation or Kessler parameterisation, ice phase	Gas phase: RADM2 Aqueous phase: uptake of SO_2, H_2O_2, O_3, etc., production of SO_4^{2-}
DESCAM Convective cloud	Model microphysical scavenging processes (Floßmann, 1994b)	2-D 20×20 km, detailed micro-physics (1 μm - 2.5 mm drop size), variable grid increments	Gas phase: none Aqueous phase: uptake of SO_2, H_2O_2, HNO_3, production of SO_4^{2-}
No Acronym Cloud layer	Study effects of clouds on ozone and photochemistry in the aqueous phase (Matthijsen, 1994)	1-D (1–3 km) turbulent transport, cloud layer 1–1.5 km monodisperse drops	Gas phase: 180 reactions, 53 species Aqueous phase: 38 reactions, 25 species.
CHEMSTAR Cap cloud	Study wet, dry and occult deposition in complex terrain (Bower et al., 1991)	Turbulent stratified air flow in complex terrain; explicit micro-physics, drop sedimentation	Gas phase: none Aqueous phase: uptake of SO_2, H_2O_2, O_3, HNO_3, NH_3, production of SO_4^{2-}
FOG Radiation fog	Study of life cycles of fogs(Bott and Carmichael, 1993)	1-D (1–3 km) higher order turbulence closure, spectral microphysics radiation effects	Gas phase: 112 reactions, 53 species Aqueous phase: 32 reactions, 23 species.
No Acronym Radiation fog	Study of life cycle of fogs(Forkel et al., 1990)	1-D (1–3 km) turbulent transport, monodisperse fog drops	Gas phase: RADM2 Aqueous phase: uptake of SO_2, H_2O_2, O_3, N_2O_5, production of SO_4^{2-}, NO_3^-
MOCCA Simulated cloud	Sensitivity studies of detailed chemistry in aqueous phase (Sander, 1994)	Box model, monodisperse drops, physical parameters according to application	Gas phase: 78 reactions, 42 species Aqueous phase: 85 reactions, 48 species Chemical scheme expandable
MOGUNTIA Convective cloud	Global scale transport model (Lelieveld and Crutzen, 1991)	3-D ($10° \times 10°$) 10 layers ($\leq$ 100 hPa) monthly averaged wind and cloud field, monodisperse drops.	Gas phase: 37 reactions, 21 species Aqueous phase: 26 reactions, 23 species Chemical scheme expandable

frequently. A possible compromise, exemplified in the global transport model used by GLOMAC, is the treatment of cloud chemistry by a subroutine in which a constant drop size is assumed. This model, however, works with an averaged cloud cover and some additional assumptions that may be unsuitable for regional transport models.

Another complication is the existence of different types of clouds that must be treated separately in models, especially convective and stratiform clouds. The first type is associated with and partly responsible for the vertical transport and mixing of trace substances in the troposphere. The second type often develops as a result of radiation imbalance associated with temperature inversion layers. Because of their different behaviour both types of clouds must be treated differently in models. The fact that most clouds evaporate rather than precipitate so that trace substances incorporated in cloud water are dissipated again must be taken into account. Thus, trace substances in the atmosphere are cycled through clouds a number of times before they are finally rained out. Including these features in models presents a considerable challenge.

The development of concepts for the incorporation of in-cloud chemistry in transport models clearly will require more effort in the future. Apart from finding efficient ways for a realistic representation of clouds in models, there also exists a need for the simplification of aqueous-phase chemical mechanisms by using lumped reaction mechanisms.

4.3.3   Atmospheric deposition

There are two key applications of our understanding of deposition processes. First, in the long range transport of pollutants throughout Europe, it is necessary to define the removal rate of pollutants at the ground to simulate adequately the trans-boundary exchange of sulfur and nitrogen. This we can summarise as contributing to the construction of matrices quantifying source contributions to the deposition to different receptors all over Europe. To satisfy this requirement, the information must be appropriate for all of the terrestrial and water surfaces, but the spatial resolution of the information is only required on a scale resolving transboundary fluxes and source regions. For the substances causing acidification and eutrophication the 50 km scale is appropriate.

For some compounds, primarily ammonia, the scale for establishing atmospheric source-receptor relationships may need to be smaller. However for most situations, transboundary fluxes of ammonia can be handled on a 50 km scale.

The other important application is the provision of input estimates for the sensitive receptors so as to provide the 'dose' in dose-effect analysis. This requirement is particularly demanding in the case of the critical-loads approach, where it is necessary to associate the spatial variability of land cover with deposition information on the same scale. In this case there is a requirement for high spatial resolution (ideally on the scale of the receptor), but it is necessary to recognise that there are important limitations to available data and our understanding of processes at that scale. The information is not necessary for all receptors since some are not sensitive to the inputs or show little or no spatial variability (for example sea water surfaces in the case of acidification).

In the following paragraphs examples will be given on the application to policy and policy-oriented issues of the deposition research within EUROTRAC.

a. Application of dry deposition estimates to terrestrial ecosystems

The development of models to provide high resolution dry deposition estimates has been the most important application of the improved understanding of surface-atmosphere exchange processes. These models have been used in the Netherlands to provide dry deposition estimates on a scale of 5×5 km (Lövblad *et al.*, 1993; Erisman *et al.*, 1993a, 1993c, 1994). Similar approaches have been used within the UK to estimate land-use-specific annual dry deposition of SO_2 (Fowler *et al.*, 1994b). These developments are currently being incorporated into schemes within

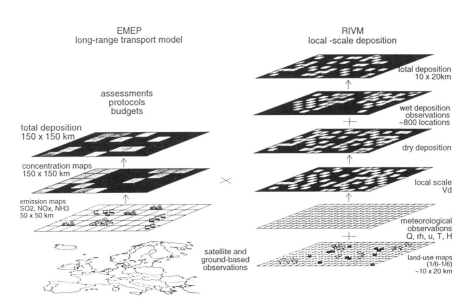

Fig. 4.4:   A scheme for calculating the fine scale deposition fluxes using a combination of nested models including those for dispersion and surface fluxes with emission, meteorology and land use data (Erisman and Baldocchi, 1994).

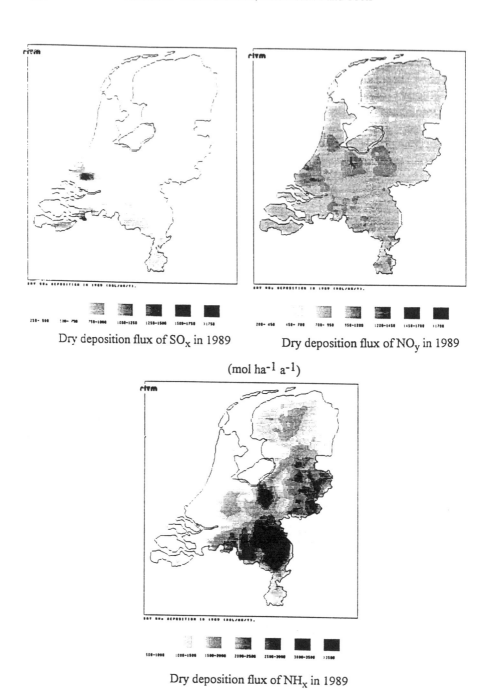

Dry deposition flux of SO_x in 1989

Dry deposition flux of NO_y in 1989

$(mol\ ha^{-1}\ a^{-1})$

Dry deposition flux of NH_x in 1989

Fig. 4.5:   Small-scale deposition maps for the Netherlands (Erisman, 1993).

the EMEP model for application across Europe (Fig. 4.4 and 4.5). Using these models, it can be shown that dry deposition of sulfur dioxide varies considerably due to receptor type and the biological and physico-chemical status of receptor and receptor surfaces. Preliminary calculations with the model show that the import and export of pollutants between countries in Europe may be altered considerably, for some countries by as much as 30 % (Erisman *et al.*, 1994).

Similar approaches to those for sulfur are being applied to NO_2 both in the UK and in the Netherlands. In the case of the application of high resolution deposition estimates (Erisman *et al.*, 1993b; UK DoE, 1994), the work has been extended to include reduced nitrogen and with a few simplifying assumptions provide the input of all sulfur and nitrogen compounds over Europe.

The exchange of NH_3 between the atmosphere and terrestrial surfaces is bi-directional, that is it is emitted to and deposited from the atmosphere according to the prevailing surface and atmospheric conditions. By studying the underlying

processes of transport and chemical and physical interactions under a range of conditions, the mechanisms of exchange over vegetation have been estimated. These have shown that the semi-natural ecosystems (for example heathlands, forest, moorland and unmanaged grassland) are very efficient absorbers of NH_3, and especially when foliar surfaces are wet with rain or dew, and that they absorb NH_3 from the atmosphere as rapidly as it is supplied to the surface by turbulent diffusion (Duyzer *et al.*, 1992). For fertilised agricultural crops the exchange of NH_3 with the atmosphere is also bi-directional. When ambient NH_3 concentrations are below the compensation point (due to ammonia production and consumption processes in plant metabolism), the vegetation fluxes are upwards from the plant canopy into the atmosphere and vice versa. These studies have been used to develop models which simulate the net exchange of reduced nitrogen (Sutton *et al.*, 1993). Such approaches are vital in the assessment of the relative importance of reduced and oxidised nitrogen to the problems of long-range transport and deposition of feed nitrogen.

The improved understanding of surface-atmosphere exchange has made it possible to estimate both the absolute amounts deposited and the relative importance of reduced (NH_x) and oxidised nitrogen (NO_y) within the total for the major land use type of the respective countries. The striking feature of the estimates of national maps of nitrogen deposition for the Netherlands (RIVM) and the UK (ITE) is the large area over which reduced nitrogen dominates the total, especially when the receptors considered are the natural and semi-natural areas. For example, in the UK the NH_3 sensitive receptors represent only 30 % of the land area yet they receive 50 % of the total nitrogen deposited in the UK (wet and dry) and of that, 70 % is in the form of reduced nitrogen ($NH_3 + NH_4^+$) (RGAR, 1990).

b. Effects of cloud droplet deposition and orography

Some of the ecosystems in Europe which are most sensitive to the inputs of acidity and nitrogen are found in nutrient poor, slowly weathering soils, often at high altitude. Such ecosystems are commonly exposed to orographic cloud, which when associated with air from polluted regions in Europe contain large concentrations of the major ions SO_4^{2-}, NO_3^-, NH_4^+ and acidity. The study of processes of cloud-droplet deposition onto moorland and forest surfaces has enabled the development of models to simulate inputs of sulfur and nitrogen to such sensitive ecosystems on a regional scale (Fowler et al., 1991; Gallagher et al., 1992; Pahl and Winkler, 1994).

The enhancement of wet deposition as a consequence of seeder-feeder scavenging of low level cloud has also formed an important part of the assessment of pollutant inputs to terrestrial surfaces in some of the high wet deposition regions of northern Europe (Fowler et al., 1993). The regions of Europe most important for these effects are predominantly western and northern areas of the British Isles and the western upland regions of Scandinavia. These areas include some of the most acid and nitrogen sensitive ecosystems of Europe in which inputs currently exceed the critical load by substantial margin and which are therefore particularly important components of the European landscape for the development of emission control strategies.

c. Estimates of total input to terrestrial ecosystems for assessing ecosystem effects and exceedances of critical loads

The application of the most recent knowledge of atmospheric surface exchange to quantify ecosystem inputs and map critical load exceedances has become the main channel for the use of BIATEX research. The methods to assess critical loads, critical levels and exceedances have mostly been developed during the 8 years of EUROTRAC. Furthermore, the research community, in designing field experiments to understand processes, has been closely involved in the development of tools to estimate inputs, and therefore exceedances. The philosophy of the critical loads approach, to bring the atmospheric inputs to ecosystems within the capacity of the ecosystem to assimilate or tolerate the input without damage, has provided a clear focus for research into the questions which limit the development of strategies for control.

The identification of specific ecosystems, especially forests and semi-natural vegetation, which are particularly sensitive, illustrates the way components of the landscape have been selected for study. The current international discussion on a nitrogen protocol which in time will lead to reductions in emission of fixed nitrogen to the atmosphere over Europe, requires an assessment of the transport, deposition and effects of current emissions. The study of processes of atmospheric-surface exchange of gaseous NO, NO_2, HNO_3 and NH_3 and particulate NO_3^- and NH_4^+ forms a vital step in the assessment. The study of the deposition processes has shown that the spatial variability in inputs is very large, and that the application of spatially uniform deposition rates, common in long range transport models, greatly underestimates the input to many sensitive parts of the landscape. This is an essential step for the protection of the nature reserves and other sensitive parts of the landscape. Current approaches to map deposition throughout Europe are limited by the availability of land-use information, but for some areas, notably the Netherlands, southern Scandinavia and the UK, rapid progress has been made in assessing the spatial variability in deposition.

It is probable that the development of a nitrogen protocol will proceed more rapidly than our understanding of inputs and effects, which illustrates the close link between the political and scientific development. Over the period of EUROTRAC the main field experiments underpinning the understanding of processes, which have fed directly into policy development based on critical loads, have been initiated within EUROTRAC and have involved research groups from different countries.

d. Input budgets of nitrogen compounds to marine ecosystems

Direct atmospheric deposition to sea surfaces is an important source of nitrogen for the primary production and thus for marine eutrophication. Budget estimates indicate that the direct deposition may contribute approximately 30 % of the total input to the North Sea (Jickells, 1993) and the Kattegat (Asman et al., 1994b). In addition, atmospheric deposition may give an indirect input due to leaching of nitrogen deposited on land surfaces. Many forested areas in the European continent leach substantial fractions of the atmospheric deposition (Malanchuck and Nilsson, 1989; Reynolds et al., 1994).

Data on the total inputs and their sources are important for the assessment of the effects and for the development of abatement strategies. Within the conventions, discussions have started on developing control strategies based on effects, that is on critical loads. In such effect-based strategies, quantification of the total atmospheric deposition to the different marine systems will be necessary.

4.3.4   The development and application of source-receptor models for policy

Overall modelling of the relationships between emissions and deposition of sulfur and nitrogen compounds for policy has mainly been done as a part of the Co-operative Programme for Evaluation and Monitoring (EMEP) under the LRTAP Convention (UN-ECE, 1994). Models have been developed and improved in order to establish quantitative source-receptor matrices. EMEP is presently obtaining yearly budgets of the transport of sulfur and nitrogen compounds over Europe. The continuous need for source-receptor data has forced EMEP to work with a top-down approach, starting with simplifications of the processes and mechanisms and then successively improving the model to incorporate improved knowledge and understanding.

For the quantitative description of the distribution and deposition of sulfur and nitrogen within Europe, Lagrangian as well as Eulerian models have been used. Within EMEP a Lagrangian model is used because of its simplicity in establishing source-receptor matrices. The Eulerian models such as EURAD, are complimentary, describing and quantifying dispersion and deposition processes not easily described by the Lagrangian models (see section 5.4).

The EMEP modelling of sulfur and nitrogen deposition has started assessing and including some of the most important recent scientific findings on deposition many of which have been achieved within EUROTRAC. The parameterisation of the dry deposition was thus recently investigated to see if including a more sophisticated parameterisation would influence total deposition as well as the source attribution procedure (Barrett, 1994). The evaluation included in particular the aerodynamic processes in the marine surface layer. This evaluation was based on results partly achieved within EUROTRAC (Joffre, 1988; Lindfors et al., 1991 and 1993). The conclusion is that the uncertainties in source attributions due to the refined description of dry deposition of gaseous nitrogen and sulfur components may be less than 5 %.

In addition to EMEP, other atmospheric transport and deposition models have been developed and tested, such as the TREND model developed by the Dutch Institute for Public Health and Environment, RIVM (van Jaarsveld and Onderdelinden, 1990; Erisman et al., 1993c). The TREND model has been used for the development of the Dutch national policy on acidification as well as for other purposes such as the policy-oriented scenario analysis presented in the so called GLOBE reports (RIVM 1992). Within EUMAC, simulations on sulfur and nitrogen deposition have been made, primarily on an episodic basis. Models for assessing the acidic deposition on the national scale have been developed in the UK and Finland. The Finnish model which was part of EUMAC has been used to assess the importance of sources in Estonia and St Petersburg for the deposition of sulfur and nitrogen in Finland (Hongisto and Joffre, 1994). The Finnish model includes a simplified cloud chemistry in order to get a better description of wet deposition. The results from the calculations indicate that the Estonian oil shale burning contribute appreciably to atmospheric concentrations of sulfur dioxide and to atmospheric deposition of sulfur in southern Finland. The UK model for the national policy on acid deposition has applied the knowledge recently developed within BIATEX in order to obtain the fine spatial resolution (10×10 km) in dry deposition and elevated cloud deposition.

Very few models have made budget calculations of the deposition of nitrogen to marine areas. In addition to the EMEP model, a yearly estimate of the deposition of nitrogen to the North Sea was made as a part of the EUMAC project. The estimates were based on episodic runs and can only be considered as a starting point. For quantification of the nitrogen deposition to the Kattegat, a model was developed and applied (Asman *et al.*, 1994b). The calculations were made as part of a programme for assessing the eutrophication of the marine areas surrounding Denmark.

The scientific research is today included in policy development by means of mechanistic models describing the transfer of pollutants from sources to receptors and sometimes also their effects. The results from the model calculations are then further linked into the process of developing policies as being part of the integrated assessment models outlined in Chapter 2.

4.3.5   *The influence of sulfate aerosols on the radiation balance*

Atmospheric aerosols influence the radiation balance by absorption and reflection of solar radiation. They may also influence the optical properties of clouds as well as the lifetime of clouds. The phenomenon and its overall sources and consequences was only discovered in 1985. In 1987 the role of phytoplankton in producing DMS and their possible importance as a climate regulating factor was first discussed (Charlson *et al.*, 1987). The importance of anthropogenic sulfur dioxide emissions for the backscattering of solar radiation was established within EUROTRAC (Charlson *et al.*, 1991). EUROTRAC subprojects ASE and GCE have, as mentioned earlier, contributed to the understanding of the processes regulating the formation and emissions of DMS as well as to the oxidation of DMS in the atmosphere. In addition EUROTRAC research has contributed to assessing the importance of sulfur emissions to the formation of sulfate aerosols with field investigations of the formation of particles in marine atmospheres as well as in cloud systems. Significant contributions to the scientific understanding of the influence of sulfur emissions on the radiation balance were made in GLOMAC. Based on the 3-dimensional tracer transport model of the troposphere (Moguntia) the global distribution of sulfur compounds has been calculated (Langner and Rodhe, 1991; Langner *et al.*, 1992). These results have been used to estimate the influence on the radiation balance and climate from the anthropogenic sulfate aerosol. The results indicate that the anthropogenic sulfate aerosol causes an annual average reduction of incoming solar radiation of approximately 1 Wm^{-2}. The effect occurs over large, mainly continental, areas in the northern hemisphere (Fig. 4.6). The negative radiative forcing from sulfate aerosols is, based on the model calculations, almost as large as the positive effect caused by increased concentrations of carbon dioxide. The calculations are made for cloudless air. In the future, the influence of natural DMS emissions as well as sulfate aerosols on cloud formation will be included.

The results from GLOMAC are continuously linked into the IPCC process. IPCC is the scientific evaluation and assessment process serving as a guide for national and international policy of greenhouse gases. The influence from anthropogenic sulfur aerosol is important for the assessment and policies for the emissions of

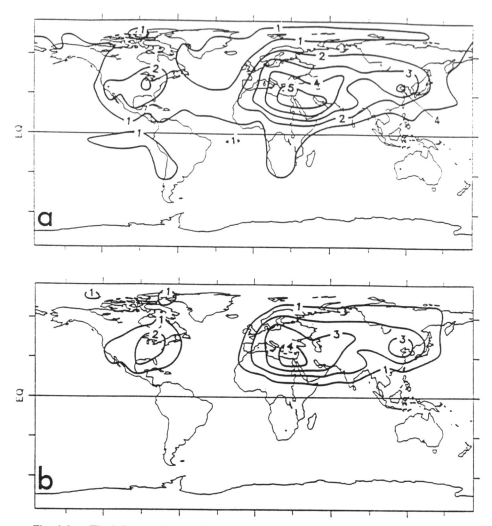

Fig. 4.6:   The influence from sulfate aerosols in the atmosphere on the global radiation
balance; a. including both anthropogenic and natural sources; b. including anthropogenic
sources only (Charlson *et al.*, 1991).

greenhouse gases. The negative effects from sulfur emissions on health and
environment will, however, require far reaching control and so decrease the
compensation effect on the climate from the sulfate aerosol. In addition, the
negative greenhouse effect from the sulfate aerosol is a short-term effect (months),
while the effect from most of the important greenhouse gases (CO_2, N_2O and
CFCs) will remain for decades and centuries.

4.4   Conclusions. Future potential for applications

4.4.1   Gaps in knowledge; uncertainties

Important steps have been taken in the development and acceptance of control strategies based on environmental effects and effect risks. The effect-oriented approach was manifested in Europe by the development and application of the critical loads concept. A prerequisite for the approach is that the actual load (the deposition) can be quantified with an accuracy and resolution in time and space of at least the same order as that of the effects, and further, that source-receptor relationships can be established with a similar accuracy. EUROTRAC has contributed to the process of developing a better scientific basis for quantifying source-receptor relationships in accordance with the overall policy questions presented in the introduction to this chapter. It has also contributed to answering the more specific questions by

- improving the basis for quantification of emissions from soils and oceans (Question 1)

- improving the parameters for modelling cloud processes (Question 2)

- establishing a detailed scheme for the quantification of deposition to various receptors (Question 3)

- providing a mechanistic basis for the interaction in dry deposition of ammonia and sulfur dioxide (Question 4)

- developing models suitable for describing the important processes for the transformation of pollutants in the atmosphere (Question 5).

An important question is therefore whether the present knowledge is adequate to set the priorities correctly and also whether the deposition goals can now be set with sufficient accuracy.

EUROTRAC was never designed to assess the overall understanding of the sources and atmospheric processes of significance for the environmental problems. Other organisations and activities are occupied with emission assessments and verifications in order to provide the best estimates of the total emissions. Such evaluations show that national inventories of sulfur and nitrogen emissions still contain large uncertainties. For nitrogen oxides and ammonia they may for certain areas be more than a factor of two. On a European scale, however, uncertainties are less. For sulfur dioxide the overall uncertainties are substantially less, primarily due to well-documented data on sulfur content in fuels and on energy statistics.

With respect to atmospheric processes, the possibility of a full evaluation of the oxidation pathways involved is extremely difficult, primarily due to the lack of data from observations and experiments, and to the difficulty in distinguishing different pathways when evaluating data from field experiments and monitoring.

The lack of a model that sufficiently quantifies the conversion of sulfur dioxide to sulfates in cloud droplets is one important shortcoming in our present understanding. Another relates to the problem of quantifying the exchange of pollutants between the atmospheric boundary layer, ABL, and the free troposphere.

The dissipation of clouds by evaporation rather than precipitation liberates aerosol particles that previously served as cloud condensation nuclei, but these are chemically altered due to in-cloud chemical processes. Each liberated particle may again act as a nucleus, so that particulate material may be cycled through clouds a number of times before it is ultimately precipitated. The effect of this process on acidification is not understood at all. In-cloud chemical processes combined with cloud dissipation certainly influence the size distribution of aerosol particles, which determines the efficiency of nucleation, and thus provides a feedback mechanism affecting both aerosol and cloud populations. Modelling as well as field studies will be required to resolve this problem.

For deposition the situation is in general simpler than for the atmospheric processes, at least for sulfur. The mechanisms and their importance can largely be verified in field experiments and by monitoring. Such verification shows that present deposition models are able to quantify the deposition to a large number of receptors, although difficulties still exist in describing the variability over deposition in complex terrain. For nitrogen there have been significant improvements in reducing the uncertainties in quantifying the input. The difficulties in verifying deposition models by measuring receptor fluxes have made it almost impossible to estimate the present uncertainties in the models with respect to the deposition of NH_3 and NO_z.

There is also a need to refine the spatial scale of deposition maps. It is clear that inputs averaged over a 150 km by 150 km grid area are not adequate for estimating the spatial distribution of effects, and ultimately data on a 1×1 km scale will be required (van Pul et al., 1995). This cannot yet be provided, even for the relatively well-understood field of sulfur deposition, and it remains a major problem for both the field measurement activity and for modelling. The ultimate need also includes the description of the sulfur and nitrogen deposition over complex terrain and to variable landscapes.

When EUROTRAC began, it was assumed that the knowledge of the atmospheric behaviour (emissions, atmospheric reactions and deposition) of sulfur was much greater than that for nitrogen compounds and photochemical oxidants. The priorities then were mainly on research on the atmospheric processes and deposition of nitrogen with less priority given to sulfur. A substantial increase in knowledge of the atmospheric processes for sulfur was, however, also achieved, especially on the wet oxidation of sulfur dioxide in droplets and on the deposition of sulfur dioxide.

4.4.2   Needs for improvements of application models

During the period of EUROTRAC, control strategies for the atmospheric pollutants distributed over Europe have changed from those being based on best available technology (BAT) and percentage reductions, towards control strategies based on the effects and effect risks also taking into account the costs of the control measures. Within the LRTAP Convention the experience gained from the sulfur protocol has resulted in a general acceptance of the concept throughout Europe that future control strategies for other pollutants should also be effect-oriented. Such approaches result in a future need for scientific developments in modelling generating quantitative source-receptor relationships, monitoring networks and assessment procedures for evaluating the effects of control measures. These approaches should provide a method to identify areas of importance for future research.

As already mentioned, the research within EUROTRAC has indicated that emissions of nitric oxide from soils may make a significant contribution to the overall NO_x emissions in Europe. The spatial variability of these emissions needs however to be further evaluated and the variations in time and space investigated. With respect to ammonia emission factors and emissions have mainly been studied in areas with intense agricultural practices (mainly the Netherlands, UK, Denmark and southern Sweden); much more work is required in less intensively cultivated areas.

With respect to atmospheric chemistry, one of the most urgent needs appears to be the incorporation in transport models of a detailed set of chemical reactions occurring in clouds, covering reactions both in gaseous and liquid phases, and the exchange of materials between them. While a number of cloud models have been developed for research purposes, they are currently too complex for incorporation in models simulating regional transport and deposition of pollutants. Even if simplifications in both the chemical schemes and the computer algorithms were made in existing computer programmes, it would be difficult to adapt them to application models. A considerable effort in the development of new and efficient algorithms will be required to deal with cloud chemical processes in models designed to predict source-receptor relationships.

Even if the full parameterisation of the deposition process for some of the most important gases were to be applied in policy-related modelling, there are still areas where further research is needed. There is also a clear requirement for long-term flux measurements (one year or more). To date, the processes have been evaluated using intensive campaigns of a few weeks measuring SO_2, O_3 and NH_3. Further development of methods is necessary to extend this to the other major gases NO, NO_2, HNO_3, and to particles.

There is a particular need to quantify the effects of complex terrain on deposition processes. The complexity takes two forms, first that due to orography, and second due to different land uses (*e.g.* forest, agriculture, urban development). In many cases these two effects interact. The key areas of uncertainty requiring further research include:

- dry deposition of NH_3, HNO_3, SO_2 and NO_2 to forests;

- cloud droplet and aerosol deposition on vegetation, especially short semi-natural vegetation and forests;

- deposition to complex landscapes, combinations of forest, open land and lakes

- deposition close to sources (10–1000 m) of NH_3, NO/NO_2 and SO_2;

- interactions between land atmosphere exchange and air chemistry in the surface layer, especially for volatile aerosols such as NH_4NO_3, NH_3 and HNO_3, for NO_2, NO and O_3 and for NH_3 aerosols and SO_2;

- long-term monitoring of dry deposition of the major gases (SO_2, NO_2 O_3, NH_3, and HNO_3) for the major land use types in Europe; forest, agricultural land and semi-natural short vegetation;

- development of the models to provide fine scale estimates of dry deposition (1×1 km) for Europe.

Regional scale modelling is presently being improved by including some of the basic work within EUROTRAC. As yet many of the achievements have not been applied in the models. This is certainly true for the cloud processes and the improved parameterisation of sulfur oxidation processes in clouds. Other areas of concern are the distribution and deposition over complex terrain. The results achieved in EUMAC have been used to illustrate some of the most difficult issues with respect to the establishment of source-receptor relationships, including distribution and deposition over complex terrain and the exchange of pollutants between the atmospheric boundary layer and the free troposphere. The inclusion of such processes into long-term policy-oriented models are still to be done.

4.5   References

Asman, W.A.H., 1992, Ammonia emissions in Europe: Updated emission and emission variations. Report 228471008, *Institute of Public Health and Environmental Protection (RIMV)*, Bilthoven, The Netherlands.

Asman, W.A.H., Berkowicz, R., Christensen, J., Hertel, O. and Runge E.H., 1994b, Atmospheric deposition of nitrogen compounds to the Kattegat (in Danish). *The Danish Marine Research Programme 90, Report No. 37*, Danish Environmental Agency, Copenhagen, Denmark.

Asman, W.A.H., Sørensen, L.L., Berkowicz, R., Granby, K., Nielsen, H., Jensen, B., Runge, E.H., Lykkelund, C., Grynning, S.H. and Sempriviva, A.M.,1994a, Dry deposition processes (in Danish). *The Danish Marine Research Programme 90, Report No. 35*, Danish Environmental Agency, Copenhagen, Denmark.

Atkinson, R., Baulch, D.L., Cox, R.A., Hampson, R.F., Kerr, J.A., and Troe, J., 1989, Evaluated kinetic and photochemical data for atmospheric chemistry; Supplement III, *J. Phys. Chem. Ref. Data* **18**, 881–1097.

Bal Reddy, K., and van Eldik, R., 1992, Kinetics and mechanism of the sulfite-induced autoxidation of Fe(II) in aqueous solution, *Atmos. Environ.* **26A**, 661–665.

Barrett, K.,1994, Dry deposition in the EMEP NOx Model: The over-sea parameterisation. *EMEP MSC-W Note 3/94*, 1–51. Oslo, Norway.

Belviso, S., Nguyen, B.C., Baut-Menard, P., Kyon, K.S. amd Rassoulzadegan, F., 1990, Production of dimethylsulfonium propionate and dimethylsulfide by the microbial food web in the Ligurian Sea. in: P.M. Borrell, P. Borrell, T. Cvitaš, W. Seiler (eds), *Proc. EUROTRAC Symp. '92*, SPB Academic Publishing bv, The Hague, pp. 79–80.

Berglund, J.S., Fronaeus, S., and Elding, L.I., 1993, Kinetics and mechanism for the manganese-catalysed oxidation of sulfur(IV) by oxygen in aqueous solution, *Inorg. Chem.* **32**, 4527–4538.

Bongartz, A., and Schurath, U.,1993, Recent determinations of mass accomodation coefficients on liquid water with an improved liquid jet technique, in: P.M. Borrell, P. Borrell, T. Cvitaš, W. Seiler (eds), *Proc. EUROTRAC Symp. '92*, SPB Academic Publishing bv, The Hague, pp. 639–643.

Bongartz, A., George, C., Kames, J., Mirabel, P., Ponche, J.L. and Schurath, U., 1994, Experimental determination of HONO mass accommodation coefficients using two different techniques, *J Atmos. Chem.* **18**, 149–169.

Bongartz, A., Schweighoefer, S., Roose, C. and Schurath, U., 1995, The mass accommodation coefficient of ammonia on water. *J Atmos. Chem.* in press.

Bott, A., and G.R. Carmichael, 1993, Multiphase chemistry in a microphysical radiation fog model - a numerical study, *Atmos. Environ.* **27A**, 503–522.

Bower, K.N., T.A. Hill, H. Coe and T.W. Choularton, 1991, Sulfur dioxide oxidation in an entraining cloud model with explicit microphysics, *Atmos. Environ.* **25A**, 2401–2418.

Buijsman, E., Maas, J.F.M. and Asman, W.A.H., 1987,. Anthropogenic NH_3 emissions in Europe. *Atmos. Environ.* **21**, 1009–1022.

Calvert, J.G. and W.R. Stockwell, 1984, The mechanism and rates of gas phase oxidations of sulfur dioxidie and the nitrogen oxides in the atmosphere. in: J.G. Calvert (ed). Acid precipitation: SO_2, NO and NO_2 oxidation mechanisms atmospheric considerations. Ann Arbor Science Publishers, Ann Arbor, Michigan, 1984.

Chang, J.S., Brost, R.A., Isaksen, I.S.A., Madronich, S., Middleton, P., Stockwell, W.R., and Walcek, C.J., 1987, A three dimensional Eulerian acid deposition model: physical concepts and formulation. *J Geophys. Res.* **92**, 14681–14700.

Charlson, R.J., Langner, J., Rodhe, H., Leovy, C.B. and Warren, S.G., 1991, Perturbation of the Northern hemispheric radiative balance by backscattering from anthropogenic sulfate aerosols. *Tellus* **43A**, 152–163.

Charlson, R.J., Lovelock, J.E., Andreae, M.O. and Warren, S.G.,1987, *Nature* **236**, 655–661.

Coe, H. and Gallagher, M.W., 1992, Measurements of Dry Deposition of Nitrogen Dioxide to a Dutch Heathland using the eddy-correlation technique. *Q. J. R. Meteorol. Soc.* **118**, 767–786.

Dore, A.J., Choularton, T.W. and Fowler, D., 1992, An improved wet deposition map of the United Kingdom incorporating the seeder-feeder effect over mountainous terrain. *Atmos. Environ.*, **26A**, 1375–1381.

Draaijers, G.P.J. and Erisman, J.W., 1993, Atmospheric sulfur deposition onto forest stands: throughfall estimates compared to estimates from inference. *Atmos. Environ.* **27A**, 43–55.

Duyzer, J. and Fowler, D., 1993, Dry deposition of nitrogen oxides, in: Models and Methods for the Quantification of Atmospheric Input to Ecosystems. *Nordiske Seminar - og Arbejdsrapporter 1993:* **573** Nordic Council of Ministers, Copenhagen, pp. 95–123.

Duyzer, J.H., Bouman, A.M.H., Diederen, H.S.M.A. and van Aalst, R.M., 1987, Measurement of dry deposition velocities of NH_3 and NH_4+ over natural terrains. *Report R 87/273.* MT-TNO, Delft, The Netherlands.

Duyzer, J.H., Deinum, G and Baak, J., 1994, The interpretation of the surface-atmosphere exchange fluxes of nitrogen oxides; corrections for chemical reactions. *Phil. Trans. Roy. Soc. A*, **350**, in press.

Duyzer, J.H., Verhagen, H.L.M., Westrate, J.H. and Boxveld, F.C., 1992, Measurement of the dry deposition flux of NH_3 on to coniferous forest. *Environ. Poll.*, **75**, 3–14.

Ebel, A., Elbern, H., Hass, H. Jakobs, J.,J., Memmesheimer, M., Laube, M., Oberreuter, A. and Piekorz, G., 1994, European acid deposition model EURAD, *EUROTRAC Annual Report part 5:EUMAC*, EUROTRAC ISS, Garmisch-Partenkirchen, pp. 26–40.

Erisman, J.W. and Baldocchi, D., 1993, Dry deposition of sulfur dioxide. In: Models and Methods for the Quantification of Atmospheric Input to Ecosystems. *Nordiske Seminar - og Arbejdsrapporter 1993:* **573**. Nordic Council of Ministers, Copenhagen, pp. 75–93.

Erisman, J.W. and Wyers, G.P., 1993, Continuous measurements of surface exchange of SO_2 and NH_3: implications for their possible interaction in the deposition process. *Atmos. Environ.* **27A**, 1937–1949.

Erisman, J.W., 1993, Acid deposition to nature areas in the Netherlands: part 1. Methods and results, *Water, Air and Soil Pollution*, **71**, 51–80.

Erisman, J.W., Beier, C., Draaijers and Lindberg, S., 1993a, Deposition monitoring. In: Models and Methods for the Quantification of Atmospheric Input to Ecosystems. *Nordiske Seminar - og Arbejdsrapporter 1993:* **573** Nordic Council of Ministers, Copenhagen, pp. 163–183.

Erisman, J.W., Jaarsweld, H. van, Pul, A. van, Fowler, D., Smith, R.I. and Lövblad, G., 1993c, Comparison between small scale and long range transport modelling. In: Models and Methods for the Quantification of Atmospheric Input to Ecosystems. *Nordiske Seminar - og Arbejdsrapporter 1993:* **573** Nordic Council of Ministers, Copenhagen, pp. 185–197.

Erisman, J.W., Potma, C., Draaijers, G., van Leeuwen, E. and van Pul, A., 1994b, A generalised description of the deposition of acidifying pollutants on a small scale in Europe. in: P.M. Borrell, P. Borrell, T. Cvitaš, W. Seiler (eds), *Proc. EUROTRAC Symp. '94*, SPB Academic Publishing bv, The Hague, pp. 588–596.

Erisman, J.W., van Elzakker, B.G. and Mennen, M.G., 1990, Dry deposition of SO_2 over grassland and heather vegetation in the Netherlands. Report No 723001004. *National Institute of Public Health and Environmental protection*, Bilthoven, The Netherlands.

Erisman, J.W., van Pul, A. and Wyers, P., 1994a, Parameterization of dry deposition mechanisms for the quantification of atmospheric input to ecosystems. *Atmos. Environ.* **28**, 2595–2607.

Erisman, J.W. and Baldocchi, 1994 Modelling dry deposition of SO_2, *Tellus*, **46B** 159–171.

Erisman, J.W., Versluis, A.H., Verplanke, T.A.J.W., de Haan, D., Anink, D., van Elzakker, B.G., Menner, M.G. and van Aalst, R.M., 1993b, Monitoring the dry deposition of SO2 in the Netherlands: results for grassland and heather vegetation. *Atmos. Environ.* **27A**: 1153–1161.

Floßmann, A.I., Cvitaš, T., Möller, D., and Mauersberger, G., 1994, *Clouds, models and mechanisms*, EUROTRAC ISS, Garmisch-Partenkirchen, 1995.

Floßmann, A.I., 1994b, A 2-D spectral model simulation of the scavenging of gaseous and particulate sulfate by a warm marine cloud, *Atmos. Res.* **32**, 233–248.

Forkel, R., Seidl, W., Dlugi, R. and Deigele, E., 1990, A one dimensional numerical model to simulate formation and balance of sulfate during radiation fog events, *J. Geophys. Res.* **95**, 18501–18515.

Fowler, D., Baldocchi, D. and Duyzer, J.H., 1991, Inputs of trace gases, particles and cloud droplets to terrestrial surfaces. In: Acidic deposition its nature and impacts. F.T. Last and R. Watling (eds), *Proc. Roy. Soc. Edinburgh* **97B**, 35–59.

Fowler, D., Gallagher, M.W. and Lovett, G.M., 1993, Fog, cloudwater and wet deposition. In: Models and Methods for the Quantification of Atmospheric Input to Ecosystems. *Nordiske Seminar - og Arbejdsrapporter 1993*: **573**. Nordic Council of Ministers, Copenhagen, pp. 51–73.

Fowler, D., Leith, I.D., Smith, R.I., Choularton, T.W., Inglis, D. and Campbell, G., 1994b, Atmospheric inputs of Acidity, Sulfur and Nitrogen in the UK. *UCL Conference on Critical Load of Acidic Deposition.* in press.

Fowler, D., McDonald, J., Leith, I.D., Hargreaves, K.J. and Martynoga, R., 1994a, The response of peat wetland methane emissions to temperature, water table and sulfate deposition. In: *Acid rain research: Do we have all the answers?* Elsevier Science, in press.

Fricke, W. and Beilke S., 1993, Changing concentrations and deposition of sulfur and nitrogen compounds in central Europe between 1980 and 1992. In: Slanina, J, Angeletti, G. and Beilke S. (eds), General assessment of biogenic emissions and deposition of nitrogen compounds, sulfur compounds and oxidants in Europe. *Joint Workshop CEC/BIATEX of EUROTRAC*, 4–7 May 1993, Aveiro Portugal, pp. 9–30.

Friedrich, R., Heymann, M. and Kasas, Y., 1994, *EUROTRAC Annual Report part 5: GENEMIS*, EUROTRAC ISS, Garmisch-Partenkirchen.

Fuzzi, S. (Editor) (1994) The Kleiner Feldberg cloud experiment 1990, *J. Atmos. Chem.* **19**, 1–258.

Gallagher, M.W., Chourlaton, T.W., Wicks, A.J., Beswick, K.M., Coe, H., Sutton, M., and Duyser, J.H., 1992, Measurements of aerosol exchange to a heather moor. in: P.M. Borrell, P. Borrell, T. Cvitaš, W. Seiler (eds), *Proc. EUROTRAC Symp. '92*, SPB Academic Publishing bv, The Hague, pp. 694–698.

Granat. L., 1993, Dry deposition of nitrogen components to coniferous forests, *EUROTRAC Annual Report part 4: BIATEX*, EUROTRAC ISS, Garmisch-Partenkirchen, pp. 122–1126

Grennfelt, P., Hov, Ø. and Derwent, R.G., 1994, Second generation abatement strategies for NO_x, NH_3, SO_2 and VOC. *Ambio* **23**, 425–433.

Hargreaves, K.J., Fowler, D., Storeton-West, R.I. and Duyzer, J.H., 1992, The exchange of nitric oxide, nitrogen dioxide and ozone between pasture and the atmosphere. *Environ. Pollut.* **75**, 53–60.

Hargreaves, K.J., Skiba, U., Fowler, D., Arah, J., Wienhold, F.G., Klemedtsson, L. and Galle, B., 1994, Measurement of nitrous oxide emission from fertilized grassland using micrometeorological techniques. *J. Geophys. Res.* **99**, 16569–16574.

Hass, H., 1991, Description of the EURAD Chemistry-Transport Model (CTM) Version 2, Mitteilungen des Instituts für Geophysik und Meteorologie, Universität Köln, no. 83.

Heintzenberg, J. (ed), 1992,The Po Valley Fog Experiment 1989, *Tellus* **44B**, 443–651.

Hobbs, P.V.(ed), 1993, *Aerosol-Cloud-Climate Interactions*, Academic Press, San Diego.

Hongisto, M. and Joffre, S.M., 1994, Transport modelling over sea areas. *EUROTRAC Annual Report part 5: GENEMIS,* EUROTRAC ISS, Garmisch-Partenkirchen, pp. 67–72.

IPCC, 1994, Climate change 1994, *The IPCC Scientific Assessment*, Intergovernmental Panel on Climate Change, Cambridge University Press.

Jickells, T., 1993, On deposition of gases and particles, *EUROTRAC Annual Report part 3 ASE*, EUROTRAC ISS, Garmisch Partenkirchen, pp. 19–21.

Joffre, S.M., 1988, Modelling the dry deposition velocity of highly soluble gases to sea surface. *Atmos. Environ.* **22**, 1137–1146.

Kramm, G., Muller, H., Fowler, D., Hofken, K.D., Meixner, F.X. and Schaller, E., 1991, A modified profile method for determining the vertical fluxes of NO, NO_2, Ozone and HNO_3 in the atmospheric surface layer. *J. Atmos. Chem.* **13**, 265–288.

Langner *et al.* 1994

Langner, J. and Rodhe, H., 1991, A global three-dimensional model of the tropospheric sulfur cycle. *J Atmos. Chem.* **13**, 255–263.

Langner, J., Rodhe, H., Crutzen, P.J. and Zimmermann, P., 1992, Anthropogenic influence on the distribution of tropospheric sulfate aerosol. *Nature* **359**, 712–716.

Larsen, S.E., Baeyens, W., Belviso, S., Buat-Menard, P., Collin, J-L., Donard, O.F.X., Harrison, R.M., Leeuw, G. de, Liss, P.S., Rapsomanikis, S., Schulz, M., 1993, *EUROTRAC Annual Report part 3 ASE*, EUROTRAC ISS, Garmisch Partenkirchen.

Leck, C., 1989, Do Marine Phytoplankton contribute to the atmospheric sulfur balance of northern Europe? PhD Thesis. Department of Meteorology, University of Stockholm.

Lelieveld, J., and Crutzen, P.J., 1991, The role of clouds in tropospheric photochemistry, *J. Atmos. Chem.* **12**, 229–267.

Lindfors, V., Joffre, S.M. and Damski, J., 1991, Determination of the wet and dry depositions of sulfur and nitrogen compounds over the Baltic Sea using actual meteorological data. *Finnish meteorological Institute Contributions* No 4, Helsinki.

Lindfors, V., Joffre, S.M. and Damski, J., 1993, Meterological variability of the wet and dry deposition of sulfur and nitrogen over the Baltic Sea.*Water, Air &Soil Poll.* **66** 1 – 28.

Lopez, A., 1994, Biosphere Atmosphere exchanges: Ozone and aerosol dry deposition velocities over a pine forest. *EUROTRAC Annual Report part 4: BIATEX*, EUROTRAC ISS, Garmisch-Partenkirchen, pp. 80–87.

Lövblad, G., Erisman, J.W. and Fowler, D. (Eds.), 1993, Models and methods for the quantification of atmospheric input to ecosystems. *Nordiske Seminar - og Arbejdsrapporter 1993*: **573**. Nordic Council of Ministers, Copenhagen.

Malanchuck, J.L. and Nilsson, J., The Role of Nitrogen in Acidification of Soils and Surface Waters, *NORD Report 1989:92*, Nordic Council of Ministers, Copenhagen.

Malin, G., Turner, S.M., Liss, P., Holligan, P and Harbour, D., 1993, Dimethyl sulfide and dimethylsulfoniopropionate in the Northeast Atlantic during the summer coccolithophore bloom. *Deep-Sea Research* **40**, 1487–1508.

Matthijsen, J., 1994, Cloud model experiments of the effect of iron and copper on tropospheric ozone under marine and continental conditions, *Met. and Atm. Phys.* in press

Meixner, F.X. and Ludwig, J., 1992,. A Field Experiment of Surface Exchange of Trace Gases: Ecosystem 'Wheatfield', Manndorf/Lower Bavaria. In: Angeletti, G., Beilke, S. and Slanina, J. (eds), *Air Pollution Research Report* **39** pp. 325–337.

Mölders, N., Hass, H., Jakobs, H.J., Laube, M. and Ebel, A., 1994, Some effects of different cloud parametrizations in a mesoscale model and a chemistry transport model, *J. Appl. Meteor.* **33**, 527–545.

MSC-E, 1993, *Annual Report.* Co-operative Programme for Monitoring and Evaluation of the Long Range Transmission of Air Pollutants in Europe. Meteorological Synthesizing Center - East, Moscow.

Müller, H., Kramm, G., Meixner, F.X., Dollard, G., Fowler, D., and Possanzini, M., 1993, Determination of HNO_3 deposition by modified Bowen-ratio and aerodynamic profile techniques, *Tellus* **45B**, 346–367.

Ottley, C.J, Harrison, R.M., 1992, The spatial distribution and particle size of some inorganic nitrogen, sulfur and chlorine species over the North Sea. *Atmos. Environ.*, **26A**, 1689–1699.

Pahl, S., and Winkler, P., 1994, Deposition von Säure und andere Luftbeimengungen durch Nebel. *Final Report BMFT, Grant 07EU726*, Hamburg.

Pahl, S., Winkler, P., Schneider, T., Arends, B., Schell, D., Maser, R. and Wobrock, W., 1994, Deposition to a coniferous forest at Kleiner Feldberg. *J Atmos. Chem.* **19**, 231–252.

Pasiuk-Bronikowska, W. and Rudzinski, K.J., 1993, Development of quasistationary kinetics of Co-catalyzed autooxidation of sulfite, in: P.M. Borrell, P. Borrell, T. Cvitaš, W. Seiler (eds), *Proc. EUROTRAC Symp. '92*, SPB Academic Publishing bv, The Hague, pp. 1026–1029.

Pasiuk-Bronikowska, W. and Rudzinski, K.J., 1994, Nonstationary autooxidation of metal ions in oxidation of sulfite, in: P.M. Borrell, P. Borrell, T. Cvitaš, W. Seiler (eds), *Proc. EUROTRAC Symp. '94*, SPB Academic Publishing bv, The Hague.

Ponche, J.L., George, C., and Mirabel, P., 1993, Mass transfer at the air/water interface: mass accomodation coefficients of SO_2, HNO_3, NO_2 and NH_3, *J. Atm. Chem.* **16**, 1–21.

Remde, A. and Conrad, R., 1991, Role of nitrification and denitrification for NO metabolism in soil. *Biogeochemistr* **12**, 189–205.

Restelli, G., and Angeletti, G., Editors, 1993, Dimethylsulfide: Oceans, Atmosphere and Climate, *Proc. Int. Symp.* Belgirate, Italy, Oct. 1992, Kluwer Academic Publishers, Dordrecht.

Reynolds, B. Ormerod, S.J. and Gee, A.S., 1994, Spatial patterns in stream nitrate concentrations in relation to catchment forest cover and forest age. *Environ. Pollut.* **84**, 27–34.

RGAR, 1990, Acid Deposition in the United Kingdom 19861988. *UK Review Group on Acid Rain. 3rd Report.* Department of the Environment, London.

RIVM, 1992, The Environment in Europe: a Global Perspective, 1–119. *National Institute of Public Health and Environmental Protection*, Bilthoven, The Netherlands.

Rondon, A., Johansson, C. & Sanhueza, E., 1993, NO emissions from soils and termite nests in a trachypogon savanna of the Orinoco basin. *J. Atmos. Chem.* **17**, 293–306.

Ruijgrok, W., Nicholson, K.W. and Davidson, C.I., 1993, Dry deposition of particles. In: Models and Methods for the Quantification of Atmospheric Input to Ecosystems. *Nordiske Seminar og Arbejdsrapporter 1993:* 573. Nordic Council of Ministers, Copenhagen, 145–161.

Sander, R.,1994, Modellierung von chemischen Vorgängen an und in Wolken, Thesis, Universität Mainz

Skiba, U., Fowler, D. & Smith, K., 1994, Emissions of NO and N_2O from soils. Environmental Monitoring and Assessment, Slanina, S. (Ed.) (1994). *EUROTRAC Annual Report part 4. BIATEX,* EUROTRAC ISS, Garmisch Partenkirchen, pp. 153–158.

Skiba, U., Hargreaves, K.J., Smith, K.A. and Fowler, D., 1992, Fluxes of nitric and nitrous oxides from agricultural soils in a cool temperate climate. *Atmos. Environ.*, **26A**, 2477–2485.

Slanina, J., Angeletti, G. and Beilke, S. 1994, (Eds.) General Assessment of Biogenic Emissions and Deposition of Nitrogen Compounds, Sulfur Compounds and Oxidants in Europe. *Air Pollution Research Report* **47**, CEC, Brussels.

Slinn, W.G.N, 1992, Predictions for particle deposition to vegetative canopies. *Atmos. Environ.*, **14** 1013–1016.

Stockwell, W.R. and J.G. Calvert, 1983, The mechanism of HO-SO2 reaction, *Atmos. Environ.* **17**, 2231 – 2235.

Sutton, M.A., Fowler, D. and Moncrieff, J.B., 1991, The dry deposition of atmospheric ammonia.   In: A. Anagnstopoulos and P. Day (eds), *Proc. 2nd Int. Conf. on Environmental Pollution*, University of Thessaloniki, Greece, pp 119–129.

Sutton, M.A., Adman, W.A.H. and Schjörring, 1993, Dry deposition of reduced nitrogen. In: Models and Methods for the Quantification of Atmospheric Input to Ecosystems. *Nordiske Seminar og Arbejdsrapporter 1993:* **573**. Nordic Council of Ministers, Copenhagen, pp. 125–143.

Sutton, M.A., Schjörring, J.K. and Wyers, J.P., 1995, Plant atmosphere exchange of ammonia. *Proc. Roy. Soc.* A, in press.

Tarrason, L., 1991, Biogenic sulfur emissions from the North Atlantic Ocean. *EMEP MSC-W Note 3/91,* The Norwegian Meteorological Institute, Oslo.

Tuovinen, J.-P., Barrett, K. and Styve, H.,1994, Transboundary Acidifying Pollution in Europe. Calculated fields and budgets 1985–94. *EMEP/MSC-W Report 1/94.* The Norwegian Meteorological Institute, Oslo.

UK DoE, 1994, *Impact of Nitrogen Deposition in Terrestrial Ecosystems.* Department of Environment, London, pp 1–110,.

UN-ECE, 1994, Protocol to the 1979 Convention on long-range Transboundary Air Pollution on Further Reduction of Sulfur Emissions. *ECE/EB.AIR/40.* Geneva.

van Jaarsveld and Onderdelinden, 1990, TREND; an analytical long-term deposition model for multi-scale purposes. Report 228603009, *National Institute for Public Health and Environment Protection*, Bilthoven, the Netherlands.

van Pul, W.A.J., Potma, C., Leeuwen, E.P., van Draaijers, G.P.J. and Erisman, J.W., 1995, EDACS: European Deposition maps of Acidifying Compounds on Small Scale. Model description and results. *RIVM Report 722401005*, Bilthoven, The Netherlands.

Warneck, P., 1991, Chemical reactions in clouds, *Fresenius Z. Anal. Chem.* **340**, 585–590.

Warneck, P.,1988, *Chemistry of the Natural Atmosphere,* Academic Press, San Diego.

Williams, E.J., Hutchinson, G.L. & Fehsenfeld, F.C., 1992, NO_x and N_2O emissions from soil. *Global Biogeochemical Cycles*, **6**, 351–388.

Wyers, G.P., Otjes, R.P. & Slanina, J., 1993, A continuous- flow denuder for the measurement of ambient concentrations and surface-exchange fluxes of ammonia. *Atmos. Environ.* **27A**, 2085.

Ziajka, J., Beer, F.M., and Warneck, P., 1994, Iron-catalysed oxidation of bisulfite aqueous solution: evidence for a free radical chain mechanism, *Atmos. Environ.* **28**, 2549–2552.

Ziegenbein, C., Ackermann, I.J., Feldman, H., Hass, H., Memmesheimenr, M. and Ebel, A., 1994, Aerosol treatment in the EURAD Model: Recent developments and first results. in: P.M. Borrell, P. Borrell, T. Cvitaš, W. Seiler (eds), *Proc. EUROTRAC Symp. '94*, SPB Academic Publishing bv, The Hague, pp. 1176–1179.

Chapter 5

The EUROTRAC Contribution to the Development of Tools for the Study of Environmentally Relevant Trace Constituents

5.1 Introduction

The "tools" in this Chapter are products which have been constructed within the framework of EUROTRAC; they form in fact the visible and tangible results of EUROTRAC. With these tools, if they are reliable and adequate, current and future scientific and policy-directed problems can be addressed and solved, which concern both photo-oxidants as well as acidification and nutrification, and which are directed to episodic situations and long-term averages, and act on urban, continental or global scales.

5.1.1 Instruments

The most obvious tools especially within EUREKA are the instruments which have been developed in EUROTRAC. These have been tested and applied in EUROTRAC projects and can be used for environmental work in the future. The development of instruments is an essential part of the research of atmospheric chemistry, and can even be considered as an obligation we have to future scientists. Instruments are not only need to assess the present state of the atmosphere but will also be used in the future to monitor the effectiveness of the abatement strategies when they are implemented. In the present projects we use instruments which have been developed in the past, and so standing on the shoulders of our predecessors; we should contribute to further developments and improvements to underpin the future.

5.1.2 Laboratory studies

Our understanding of reaction rates, photolysis rates and reaction products is based on laboratory studies, and the scientific advances and results obtained in the laboratory can also be considered as tools.

Laboratory studies also include the development of laboratory techniques and instrumentation which are used to make measurements in the laboratory. These new instruments and techniques currently in use in the laboratory are often developed further into instruments which can be used in the field. Nearly all, perhaps all, field instruments originate from instruments first developed and tested in the laboratory.

Laboratory studies have played an essential role in the major discoveries of atmospheric chemistry, and especially the condensed form of chemical schemes within simulation models are essential to an adequate description, both scientific and policy-oriented, of the relationship between emissions and concentrations or deposition.

5.1.3 Models

A third category of tools is provided by simulation models.

Simulation models are intended to solve the continuity equation which, in a mathematical form, describes the relationship between emissions and concentrations or deposition. Several of these system-oriented simulation models have been developed within EUROTRAC. As well as their required input such as emissions and meteorological information, simulation models contain modules or sub-models which describe the processes contained in the continuity equation. For example, the chemical mechanisms or cloud models are sub-models. In general, these process-oriented models do not describe the relationship between emissions and concentrations or deposition, but describe processes. These process-oriented models themselves, like simulation models, are also tools.

Simulation models, because they provide a framework for integrating process-oriented studies, play an essential role in the transfer of knowledge between science and policy.

5.1.4 Further tools

Data bases from field experiments, and the emission and land-use data bases are also tools developed in EUROTRAC. The set-up and lay-out of an adequate and optimal measuring network of specific compounds, leading to long-term monitoring is also a tool.

Finally, and perhaps most importantly, a network of scientists has been established in EUROTRAC. The expert groups that have been formed are part of the infrastructure addressing environmental science and policy in Europe. This network tool is invaluable, and could and perhaps should be considered as the major outcome of the project.

Tools are worthless if they are of poor quality. If their quality is not adequate, they could lead to scientific mistakes and to false answers to policy-oriented questions.

Such false answers would subsequently lead to large economic losses. The quality assurance/quality control of EUROTRAC tools is primarily based on the normal process of publishing the results in scientific peer-reviewed papers and at conferences. In addition, the EUROTRAC Scientific Steering Committee performed regular reviews of the EUROTRAC subprojects to monitor their work and encourage their quality In this way an attempt has been made in EUROTRAC to create adequate tools for current and future research.

Finally, it should be stated that the tools mentioned here developed in the framework of EUROTRAC could not have been made without the already existing and ongoing research in the scientific community throughout the world.

5.2   Instruments

5.2.1   Instrument development within EUROTRAC

The development and improvement of instruments and techniques is a necessary and continuous adjunct to scientific work. In atmospheric chemistry, where instrumentation is required both for monitoring the behaviour of the atmosphere and for research into atmospheric processes, there was, and indeed still is, a particular need for sensitive, specific and fast response instrumentation.

At the beginning of EUROTRAC, in order to meet the aim of developing sensitive, specific and fast-response instrumentation, three subprojects were set up to exploit three newer and promising techniques for measuring concentrations of trace substances in the atmosphere. The initial objective was not only to stimulate the development and the assessment of new instrumentation at a European level, but also to use it in field campaigns within EUROTRAC or in ground-based networks. It was intended that the development should not only provide instruments suitable for field operation by scientists, experienced with the particular technique, but also to produce reliable instruments suitable for operation by other scientists or by skilled technical staff. In addition it was hoped that commercial instruments would also be developed from the work. The aim was a direct consequence of EUROTRAC being a EUREKA project.

Instrument development however was not just confined to the three instrument subprojects. Instruments have also been developed in the subprojects devoted to field measurements: ALPTRAC, ASE, BIATEX, GCE, TOR and TRACT. Standard instruments were improved and new ones constructed to meet specific measurement needs.

In this section the various EUROTRAC endeavours in the field of instrument development for field measurements are briefly described. Improvements in laboratory instrumentation (LACTOZ and HALIPP) are outlined in section 5.3.5.

The section concludes with some reflections on instrument development within a scientific project.

5.2.2 Instrument subprojects within EUROTRAC

Spectroscopic methods for the analysis of atmospheric constituents, such as those chosen for development in the three EUROTRAC instrument subprojects, are rapid, non-intrusive and collect the data in ways which lend themselves to rapid computer processing. Also in some configurations it is possible to measure concentrations of several species simultaneously.

The three techniques chosen were not entirely new and instruments using each had already been employed in field measurements. The challenge was to bring them to a degree of maturity so that the instruments could be used by scientists, other than those directly involved in their development, or operated by skilled technical staff. The sensitivity, selectivity and response times were to be improved to increase the range of application of the techniques and facilitate the detection of tropospheric species in the ppt and sub-ppt ranges with a time resolution of a few seconds.

An advantage offered by co-ordinated work is the possibility not only to develop instruments but also to test them in intercomparison campaigns and so explore and verify their characteristics and reliability. The groups throughout EUROTRAC have made full use of this facility.

a. Long-path optical absorption techniques: the subproject TOPAS

In the technique of long-path differential absorption spectroscopy (DOAS), a beam of light from a bright source is projected through the atmosphere for typically a kilometre or more to a receiver consisting of a spectrometer and photo-detector. With the long path length, absorptions in the visible or UV region of the spectrum, due to trace amounts of pollutants, can be detected. It is also possible to use light from natural sources such as the sun, the moon or the daylight sky. Difficulties include overlapping spectra and retrieving the concentrations from the observations but these can be solved with suitable numerical algorithms. The advantage of DOAS is the possibility of continuous simultaneous monitoring of a number of pollutants.

At the outset of the subproject TOPAS, several DOAS instruments were already in existence. These were essentially laboratory prototypes and not easily transportable. The subproject was established to develop high performance instruments based on DOAS, in order to measure tropospheric constituents using their absorption properties at UV, visible and near-IR wavelengths. Among the groups involved, three had already undertaken preliminary industrial development of DOAS systems for pollution monitoring and established commercial companies: ATMOS in France, Hoffmann Meßtechnik in Germany and OPSIS in Sweden.

Among the instrumental improvements accomplished within the subproject are the use of a Fourier transform spectrometry in DOAS measurements (Colin *et al.*, 1991), reliable long path absorption cells, automation and new algorithms for the retrieval of species concentrations (Simon *et al.*, 1994). The number of species which can be measured and detected using the DOAS technique has also increased and includes not only the examples in Table 5.1 but also NH_3, O_3, toluene, benzene, *p*-xylene, naphthalene and other aromatic compounds (Platt, 1994a).

Table 5.1:   Example sensitivities for estimation with DOAS.

Compound	Wavelength range (nm)	Detection limit / ppt (5 km light path)
SO_2	290–310	17
CS_2	320–340	500
NO	200–230	240
NO_2	330–500	80
NO_3	600–670	2
HNO_2	330–380	40
Formaldehyde	300–360	400

Platt, 1994a

An objective of TOPAS was to determine concentrations of the free radicals, OH and NO_3. Concentrations of the nitrate radical (NO_3) are now regularly measured in field campaigns (Simon *et al.*, 1994), but the measurement of OH concentrations is still a specialised activity. (Platt, 1994a)

An intercomparison campaign for DOAS instruments was organised in 1992 in Brussels. It involved eight research groups from five nations and included instruments from three commercial companies. The discrepancies observed between the instruments originated from differences in spectral resolution and the data treatment algorithms being used with the various equipment. The campaign indicated the need to establish a common data base of absorption cross-sections with a high spectral resolution, and some measurements were subsequently made using high resolution Fourier transform spectroscopy (Carleer *et al.*, 1994). More such measurements are needed. A second intercomparison was held in the UK in 1994.

DOAS instruments are in routine use by TOPAS groups in Germany and the UK and DOAS instruments have been used by TOPAS participants in a number of field campaigns in both Europe and North America.

Commercial development: the two companies which participated at the beginning of the project certainly gained expertise from the co-ordinated work, as did the three companies which took part in the intercomparison campaign. Commercial DOAS based systems are now established in several cities for pollution monitoring, and the assessment of the instrumentation has been made by environmental agencies.

b. Lidar techniques for vertical ozone soundings: the subproject TESLAS

The differential absorption lidar technique (DIAL) offers a way of monitoring the vertical concentration profile of ozone from an altitude of about a hundred metres up into the stratosphere, with a vertical resolution of a hundred meters or so in the troposphere. It is based on the measurement, as a function of time, of the light scattered back from two vertically projected and simultaneously emitted laser beams, each having a different wavelength. The light from one is partially absorbed by ozone; the other, in which there is no absorption by ozone, is used as a reference. The technique allows processes such as the transport of ozone between the planetary boundary layer and the free troposphere to be observed which would be impossible with ground based instruments. DIAL has the advantage over balloon-borne ozone sondes that it can be run almost continuously; it has the disadvantage that the altitude which can be probed by the two laser beams is limited in cloudy or foggy conditions.

At the outset of TESLAS, lidar instruments were in use for measurements, from both fixed and mobile platforms, of ozone and aerosol vertical distributions in the stratosphere and the troposphere. However the use of these instruments in field campaigns required a major effort and they could not be used in monitoring networks. The main objective of the subproject was to provide reliable DIAL instrumentation for routine measurements of ozone vertical distributions in the troposphere, and for detailed studies of atmospheric processes involving transport, production and destruction of ozone.

The technical improvements include (Papayannis et al., 1991; Kempfer et al., 1994; Bösenberg, 1994):

- the use of stimulated Raman scattering to provide the additional wavelengths for differential absorption measurements;

- the use of grating polychromators for spectral resolution of the signals received;

- laboratory studies of ultraviolet spectra of interfering species (SO_2, H_2O);

- the development of new data acquisition electronics;

- development of algorithms for data analysis;

- the semi-automation of the equipment;

- reduction in size of the apparatus to facilitate its use in aircraft.

Of these the use of Raman shifting to provide a reliable light source at the required wavelengths represents an appreciable breakthrough, particularly when powerful laser sources are used to reduce the integration time for the measurements. European scientists now occupy a leading position in this field.

An intercomparison campaign was held in Bilthoven in 1991 (TROLIX) and an algorithm intercomparison undertaken in 1993. These have led to a reassessment of existing techniques and resulted in narrowing the ranges of uncertainty (Bösenberg, 1994 and earlier annual reports).

Instruments have been used for field work too with several TESLAS instruments in regular use at TOR stations (Ancellet *et al.*, 1991; Carnuth, 1992; Trickl, 1994; Bösenberg, 1994; Sunesson *et al.*, 1994). They have also been employed in field campaigns such as TRACT and POLLUMET.

The sensitivity, height and time resolution of the technique is illustrated in Table 5.2 The variation with height is due to the weakening of the signal received with altitude.

Table 5.2:   Typical performance of a ground based tropospheric ozone lidars. (The specifications cannot be reached in regions of strong aerosol gradients such as the top of planetary boundary layer)

height range (m)	height resolution (m)	time resolution (min)	accuracy ($\mu g\ m^{-3}$)
100–1000	75	0.5	4
1000–3000	150	< 5	4
3000–10000	500–1000	< 15	8

Bösenberg, 1995

One hoped-for development was the development of lidar systems to determine the profile of water vapour concentrations in the atmosphere. It did not occur principally because of lack of funding,

Commercial development: several groups have obtained patents at the subsystem level (optics, electronics) but no commercial development of the complete lidar system is foreseen, as the market for such instruments is small and much expertise is still required in processing and interpreting the results.

c. Tunable diode-laser spectroscopy: the subproject JETDLAG

Most of the trace compounds of interest in the troposphere have absorption features in the IR wavelength range, so that IR absorption spectroscopy is a recognised tool for the monitoring of atmospheric constituents. The development of narrow linewidth solid state lasers led to the development of a new field: tunable diode laser absorption spectroscopy (TDLAS). The narrow linewidth source removes the need for a monochromator; a specific absorption line of the compound of interest can be selected from a complex spectrum and the concentration determined. The lasers can be tuned over a limited wavelength range, or several lasers used within the same instrument, to permit the determination of several components. Difficulties are in the low pressure needed to obtain sharp spectra, which can cause sampling problems, the low temperature required for laser operation and the relatively weak absorption in the IR compared say to visible and UV regions of the spectrum.

TDLAS instruments were already available commercially at the outset of the subproject, although at an early stage of development. A major objective was to increase the sensitivity of the technique and to develop fast-response instruments so that they could be used for example in eddy-correlation work for the measurement of fluxes.

A persistent problem with TDLAS is the reliability of the laser diodes, which are available from only two companies, and which constitute the active part of the system. Much of the work performed in the first years of the subproject was devoted to the development and assessment of the TDLAS components and techniques. JETDLAG work demonstrated the problems in focusing the beam from the early multimode lasers and the consequent power losses. The development of heterostructure laser architecture using PbSnTe or PbEuSeTe active layers has resulted in improved mode quality and the ability to operate with liquid nitrogen cooling, which is cheaper and eases logistical problems during field work (Brassington, 1995).

Other activities have included exploring new methods for improving the sensitivity, including photoacoustic methods, and the use of very high frequency modulation. Various methods of multiplexing have also been tried so that several compounds my be determined at the same time (Brassington *et al.*, 1994).

An indication of the sensitivity of the technique is given in Table 5.3 where some typical compounds are shown. The feasibility of using TDLAS to measure concentrations of the free radical, HO_2, has been explored but this seems some way off yet.

Table 5.3: Typical detection limits for TDLAS measurements.

Compound	Wavelength (μm)	Detection limit (ppt)
O_3	9.50	260
CO	4.65	73
N_2O	6.25	50
H_2O_2	7.79	320
formaldehyde	3.56	160

Werle and Slemr (1995)

Individual instruments have been involved in field campaigns with GCE and TRACT and on ship cruises, where in one study an instrument was operated largely without interruption over a four week period. In addition instruments have been used for eddy-correlation work and are being prepared for use in aircraft (Brassington *et al.*, 1994 and earlier annual reports).

Spectral data is a necessary prerequisite for further applications of TDLAS and high resolution spectroscopic measurements have been made for NH_3, NO, NO_2, N_2O, HNO_2, HNO_3, C_2H_2, C_2H_6, CH_3CHO and CH_3Cl (Brassington *et al.*, 1994 and earlier annual reports).

A field intercomparison campaign for TDLAS instruments is scheduled for 1995.

On the commercial side there has been contact with firms which supply TDLAS instruments; the contribution to the improvement of the lasers themselves has already been mentioned.

5.2.3   Instrument developments in the field subprojects

Instrument development took place not only in the three "instrument" subprojects but also in those concerned with field measurements. Some highlights are given here.

a. Commercial instrument development in the subproject TOR

Several approaches for automating VOC measurements have been made and implemented at the TOR stations. A prototype for automatic sampling of VOCs was built to given specifications by Chrompack, the necessary parameters for optimal separation being determined in subproject work. The instrument is used at a number of TOR stations as well as in other subprojects (Mowrer and Lindskog, 1991). An intercomparison of VOC analysis, conducted in TOR, demonstrated that a reasonable accuracy could be obtained for the more volatile species with careful work (J. Hahn in Kley *et al.*, 1994).

For the nitrogen oxides, a sensitive and accurate NO_x analyser has resulted from subproject work (ECO Physics Model CLD 770). It has been employed at several TOR stations and has been sold throughout the world (A. Volz-Thomas in Cvitaš and Kley, 1994).

Automated instruments for the measurement of peroxides and formaldehyde have been developed and are available from AEROLASER GmbH (Slemr *et al.*, 1994).

A photo-sensor for the measurement of the photolysis rate of NO_2 (Junkermann *et al.*, 1989) was improved for field use and in aircraft. It is now commercially available through Meteorologie Consult GmbH. A similar instrument for the photolysis rate of ozone, which is the primary process in formation of OH radicals is under commercial development (Volz-Thomas, 1995).

A sonde is also being developed at the KFA Jülich for the measurement of hydrogen peroxide (patent obtained, licence contract with the UNISEARCH firm).

An intercomparison of peroxy radical chemical amplifiers was held within TOR (Volz-Thomas *et al.*, 1993).

In addition the instrument developments within TOR have greatly improved the measurement capabilities in other projects (OCTA, FIELDVOC, STREAMfield, TOASTE).

b. Instrument work in the subproject BIATEX

A small, lightweight (1.5 kg), fast-response ozone sensor for direct eddy correlation has been constructed. A similar system for the measurement of total sulfur using a fast-response flame photometer is also under development (Güsten *et al.*, 1992).

A novel instrument, under development, based on the photo-thermal deflection (PTD) technique, offers the possibility of measuring fluxes of ethene and ammonia in eddy-correlation work. A closed cell is not needed and so the problem created by the absorption of polar gases on the walls of the cell is avoided. A prototype has been built and used in field measurements (De Vries, 1994).

An accurate and sensitive wet denuder system has been marketed in collaboration with the ANASYS firm. The instruments have been used in many parts of Europe to derive concentration fields and fluxes (Wyers *et al.*, 1993).

A laboratory prototype of an aerosol sampler has been constructed. Aerosols are collected by addition of steam and the formation of droplets; the solution produced in the steam jet aerosol collector is then analysed with a gas-diffusion conductivity detector (Slanina, 1995).

An automated GC system for measuring biogenically emitted compounds has been developed. It has been used to assess biogenic VOC fluxes by means of micrometeorological methods and so validate the customary enclosure methods. (Torres, 1991)

Instrument intercomparisons held during BIATEX field campaigns in Leende, Halvergate, Manndorf and in the Bayerische Wald have helped to establish the limits of accuracy for concentration and flux measurements, thus increasing the confidence in regional and global estimates of deposition and emission derived from such work (Slanina *et al.*, 1994 and earlier Annual Reports).

c. DMS precursors in the subproject ASE

Analytical methods have been developed for the analysis of dimethylsulfopropionate (DMSP) and dimethylsulfoxide (DMSO), biochemical precursors of dimethylsulfide (DMS). The deployment of reliable instrumentation in ship-borne campaigns, such as that on the Polarstern, are an important outcome of the instrument assessments conducted in this subproject (Larsen *et al.*, 1994).

d. Cloud instrumentation in the subproject GCE

The field campaigns in the Po Valley, at Great Dun Fell and on Kleiner Feldberg have been used for instrument assessment and intercomparison, particularly for improved collectors for dry and wet deposition, aerosols and precipitation. A new droplet/aerosol analyser, to study the activation of particles as a function of particle size and composition, has been developed and used in the GCE work (Schell *et al.*, 1992, Fuzzi *et al.*, 1994).

e. Intercomparison of chemical analysis in the subproject ALPTRAC

An intercomparison of analysis methods for snow samples was conducted in Davos in 1993. Satisfactory results were obtained for the ions Cl^-, NO_2^-, SO_4^{2-}, NH_4^+, K^+, Mg^{2+} and H_3O^+, but there were difficulties with common ions as Na^+ and Ca^{2+}, due to interferences, possibly from Saharan dust. Such intercomparisons are essential if one is to have confidence in results obtained from measurements made at widely separated sites. (Schikowski *et al.*, 1994)

f. Quality assurance in the subproject TRACT

Quality assurance and quality control were important parts of the TRACT campaign: a quality control instrument caravan was established for calibration of aircraft instrumentation (Mohnen *et al.*, 1993). A comparison of gas sensors in use for ground based measurements for example showed that sensors for CO and SO_2 measurements were within 5 % of each other, 10 % for ozone and 20 % for NO_x (Fiedler *et al.*, 1994).

g. New European testing facilities for the subprojects TOR and BIATEX

Finally, two facilities have been established which are available to European research groups: an ozone ECC sonde calibration facility at the KFA Jülich, developed as part of TOR (Kley *et al.*, 1994). The results have been used by the WMO in their assessment of ozone trends. A cloud wind tunnel at ECN in the Netherlands, developed as part of BIATEX, is used for calibrating instruments measuring aerosol and droplet-size distributions (Slanina *et al.*, 1994).

5.2.4 Some reflections on instrument development in EUROTRAC

A new feature in EUROTRAC, directly related to its being a EUREKA project, was the formation of three subprojects devoted specifically to instrument development. A brief balance sheet is presented here of the positive and negative aspects of instrument development within EUROTRAC.

a. Positive aspects of instrument subprojects within EUROTRAC

- EUROTRAC has played an important role in encouraging the formation of European groups, expert in the various techniques. This is now recognised at the international level, and is a large change compared with the situation ten years ago.

- The three subprojects have developed modern instruments which are sensitive, specific and have a fast response.

- There has been an overall improvement in instrument reliability which is essential for deployment on a regular basis in the field. For some instruments, the improvement has also led to the development of transportable and/or airborne versions of the equipment.

- A major contribution of the co-operative work has been the organisation of intercomparison campaigns and field validation experiments, concerned with both the methodologies (algorithms, input data *etc.*) and the measurements themselves. These exercises, performed on an European basis, have increased confidence in the techniques to a level where the instruments can now be used in field campaigns and monitoring networks with a proper appreciation of their potentialities and relevant uncertainties.

- Common data bases, for spectroscopic data and other ancillary parameters required for signal inversion *etc.*, have been developed and European facilities for ozone-sonde calibration and for testing aerosol and cloud collectors established.

- The instruments have been used extensively in experimental field campaigns.

b. Some shortcomings of the EUROTRAC instrumental subprojects

- Progress has been slow; the strong co-ordination with central funds, required for the development of commercial instruments, cannot be provided within the present structure of EUROTRAC. Furthermore the differences in the levels and timing of the funds provided by the various countries particularly hinders development work. The problems are reflected in the fact that it was necessary to reorganise both JETDLAG and TOPAS during the course of the project, and for them to continue with more limited objectives.

- While a number of instruments have been automated and commercialised, many are still at the level of research instruments or laboratory prototypes.

- Apart from the examples mentioned, there has been little actual commercial interest in the development of equipment for monitoring and research, perhaps because of the small and uncertain markets. While the techniques chosen for development are state-of-the-art technology, they are also complex and usually require expertise with the technique itself to produce reliable results. The larger scale monitoring market requires equipment that can be operated by technical staff and produces results without too much detailed scientific interpretation being required.

5.2.5   Conclusions and recommendations

- The instrumental subprojects, the science oriented experimental subprojects and the laboratory measurement subprojects have developed numerous innovative and reliable instruments. Most of these have undergone thorough validation through intercomparison exercises, and have been used in field experiments or implemented in scientific monitoring networks as part of EUROTRAC and other environmental projects within Europe. They have also led to a large increase in the international role of European research groups in measurement related activities, such as large-scale field campaigns or tropospheric networks.

- Should the specific development of commercial instruments be desired in a future project, then subprojects can only be developed efficiently if centralised funds are available to ensure the timeliness of the various contributions. More direct involvement of commercial companies contributing their own funds would be desirable.

- The development and improvement of instrumentation should certainly be encouraged, as part of the normal work in the field and laboratory subprojects in a future project. Work in EUROTRAC clearly demonstrates that this route is particularly appropriate if the validation and quality control phases of the instrument development are considered as the primary requirement.

5.3   Laboratory studies

5.3.1   Laboratory studies and their scope

a. The nature of laboratory studies

In a project devoted to the study of chemical transformation of trace substances in the atmosphere, laboratory measurements are needed to explore new possibilities of chemical reactions and physical processes which might need to be included in chemical simulation models (section 5.4), and to provide new or improved input data on reactions for such models. The outcome of the work allows parts of chemical schemes that had hitherto been speculative or uncertain to be confirmed or to be firmly rejected. New information may also suggest ways in which the schemes for use in the models can be *reduced* or *condensed* in order to simplify them (section 5.3.4).

Within EUROTRAC, the laboratory studies carried out fall into two broad categories: (i) chemical and physical investigations of systems that are of potential atmospheric importance; and (ii) the development within the laboratory of techniques either for the study of the atmosphere itself or intended to enhance the investigations of category (i).

In this section, a brief overview of the work done will be given in order to illustrate the contribution which has been made to the understanding in this area. A full scientific review is given in the other volumes published in this series.

b. Laboratory programmes within EUROTRAC

Within EUROTRAC, two subprojects were established as laboratory programmes: LACTOZ and HALIPP, which, broadly speaking, were directed at the investigation of homogeneous gas-phase chemistry on the one hand, and heterogeneous droplet and aerosol chemistry on the other. In addition, several other subprojects within EUROTRAC have substantial components of laboratory research but these will not be considered further here.

The differences in the state of knowledge between the areas covered by the two laboratory subprojects provide an interesting parallel to the stages proposed in section 2.3 for the development of environmental problems, and illustrated in Fig. 2.3. In gas-phase chemistry (LACTOZ) the major reaction pathways for normal hydrocarbons were largely clear and understood; one was at the stage of *common understanding,* although much quantitative work was still required as well as exploratory work on the oxidation of aromatic and biogenic hydrocarbons, and on the photolysis of carbonyls. The situation in heterogeneous chemistry was much less clear and still in the *discovery* phase; the mechanisms were only speculative and much of the work was still far from the quantitative detail which will ultimately be required for both understanding and modelling. The work within HALIPP has moved it closer towards the required common understanding.

5.3.2   Some highlights of the scientific advances concerned with photo-oxidants

As indicated in section 3.2.3, ozone in the lower atmosphere is produced photochemically from NO_2, which is generated by the oxidation of NO, of which combustion is a major source. The conversion of NO to NO_2 depends on a cyclic chain process in which organic peroxy radicals, RO_2, play a key role. Peroxy radicals are themselves generated from hydrocarbons and other organic compounds (volatile organic compounds, VOCs) by a sequence of reactions started by the attack of yet other radicals (OH and NO_3).

At the outset of EUROTRAC, the main elements of ozone generation were well understood, but many details were missing. The principle aim of LACTOZ was to study the chemistry of VOCs and of the nitrogen-containing inorganic and organic compounds as well as to measure the photochemical parameters of selected compounds thought to be important in the oxidation of VOCs. Table 5.4 lists the areas of study undertaken, their present status and where one can expect the results to be applied. In the following sections, selected aspects of the work are discussed in more detail. A comprehensive list of the rate constants and reaction parameters is given in the annual reports (Le Bras *et al.*, 1994 and earlier annual reports).

In addition to the individual detailed studies, the LACTOZ steering group has produced definitive reactions schemes for day and night-time oxidation of VOCs and for NOy chemistry (Le Bras et al., 1994) that indicate the reactions which should be included in chemical oxidation models.

a. Free radical and other reactions

Much work centred on key free radical intermediates in the oxidation chains. Free radicals are short-lived species and usually require sophisticated experimental techniques for their study (section 5.3.5).

During the daylight hours oxidation is known to be initiated by the hydroxyl radical (OH) which is produced photochemically (section 3.2.3). It has now been demonstrated that oxidation of VOCs can also take place at night: the reaction of ozone with NO_2 forms a nitrate radical (NO_3), that can act as an initiator for an oxidation chain reaction. (Wayne *et al.*, 1991; Moortgat, 1994).

Alkenes are not only attacked by OH but react with ozone itself, undergoing ozonolysis and often forming complex bi-functional organic compounds. These themselves are secondary pollutants and also undergo further reactions in the atmosphere. (Le Bras *et al.*, 1994; Zabel, 1994).

Similarly, the kinetics of peroxy radicals (RO_2) and alkoxy radicals (RO) have been examined since their various reactions within the oxidation of VOCs strongly influence the yields of photo-oxidants in the atmosphere (Lightfoot *et al.*, 1992; Zellner, 1993; Le Bras *et al.*, 1993; Eberhart *et al.*, 1994) and hence the so-called ozone forming potentials of the VOCs concerned.

Table 5.4:   Scientific Advances within LACTOZ

Topic	Status	Application
Alkylperoxy (RO$_2$) Hydroperoxy (HO$_2$) radicals	Much progress for simpler RO$_2$; more work needed for substituted and larger RO$_2$ radicals	Modelling of photo oxidants in regions of medium and low NO$_x$
Alkoxy (RO) radicals	Up to C$_3$ finished; some results for C$_5$; more data needed for larger compounds.	Definition of ozone production potentials
Nitrates and peroxy-nitrates (RO$_2$NO$_2$)	Thermal lifetimes finished; data for "S" instead of "C" needed. Photolysis and OH rate constants in progress.	Broad data base for modelling photo-oxidant chemistry; long-range transport of NO$_y$
Nitrate (NO$_3$) radical	Kinetic data for VOC complete; data for larger peroxy radicals needed with more product studies.	Modelling of night-time chemistry
Hydroperoxides (ROOH)	Formation from RO$_2$ and HO$_2$ finished for small RO$_2$; formation from biogenic VOCs needed. Sources from ozonolysis need exploring.	Modelling in medium to low NO$_x$ regions; distinction between O$_3$ and OH reactions with alkenes, including in aqueous phase
Ozonolysis	Recent findings have revealed many uncertainties; reliable data needed.	Necessary for modelling the impact of biogenic alkenes on photo-oxidant chemistry
Biogenic VOCs	OH with isoprene finished; terpenes and ozone reactions need further study	Necessary for modelling the impact of biogenic alkenes on photo-oxidant chemistry
Aromatic VOCs	Kinetic data for primary processes established; further work needed on ring opening; impact of NO$_x$ unclear.	Necessary for modelling photo-oxidant chemistry in urban and neighbouring areas
Photochemistry: pathways, quantum yields and photolysis frequencies	NO$_2$ and HCHO established; O$_3$ uncertain by a factor of 2; oxygenated compounds need more work; some progress with PAN	Necessary for chemical mechanisms required for modelling, in particular for biogenic and aromatic VOCs
Formation of HONO	Some progress but heterogeneous reaction needs study.	Necessary for modelling photo-oxidant chemistry in urban areas

Becker, 1995

New measurements of the rate constants for attack of OH on a number of compounds of potential atmospheric interest have been made. Apart from the possible inclusion of such reactions in the chemical models, such rate constants give an indication of the lifetime of the particular trace compound in the atmosphere and, together with other factors, allow their possible importance as pollutants to be judged (Le Bras *et al.*, 1994; Atkinson, 1994; Peters *et al.*, 1994). In addition studies have been made of the more difficult problem of OH attack on oxidation intermediates such as poly-functional organic compounds. (Bierbach *et al.*, 1994; Neavyn *et al.*, 1994)

Peroxyacetylnitrate (PAN), a pollutant which is formed in oxidation reactions of VOCs, is formed from a peroxy radical and NO_2 at low temperatures. It can also act as a temporary carrier for NO_2, allowing its transport to remoter regions, where the PAN may dissociate liberating NO_2, and thus forming photo-oxidants far from the NO_x source. There has been work on the chemical stability of PAN itself, as well as on other possible PAN-like transport agents (Le Bras *et al.*, 1994; Zabel, 1995).

b. More complex oxidation reactions

Early work on oxidation (section 2.2.3) concentrated on the chemistry of the saturated hydrocarbons (alkanes):saturated hydrocarbons (alkanes) that were both common in the atmosphere and relatively simple to study. However from the outset of LACTOZ, work has been done on the oxidation of aromatic hydrocarbons which are also important constituents of fuels, and also on naturally produced hydrocarbons, such as isoprene and terpenes. Such biogenic hydrocarbons can play an important role in determining whether a NO_x or VOC control strategy should be adopted in regions with appreciable biogenic emissions (section 2.3.2). The products of the initial attack and the subsequent reaction sequence have proved to be extremely difficult to characterise but substantial progress has been made and outline reaction schemes have been prepared for both aromatics and for isoprene (Le Bras *et al.* 1994; Becker, 1994; Hayman, 1994; Jenkin *et al.*, 1994).

c. Photochemical and spectroscopic studies

Atmospheric oxidation is driven by absorption of sunlight which both produces the OH radicals to initiate the reaction chains and also ozone which, apart from being regarded as a pollutant, is also a necessary participant in the oxidation reaction sequences (section 2.2.3). In addition some of the intermediates produced in the reaction chains also absorb sunlight and dissociate, so a knowledge of photochemical quantum yields and spectroscopic parameters for the various compounds is essential if the simulation models are to be reliable. This type of work is difficult and time consuming, and world-wide there are few research groups that undertake it but appreciable new contributions have been made, both to the data available, and to the understanding of these processes (Barnes *et al.*, 1993; Ball *et al.*, 1994; Platt *et al.*, 1994a,b; Brion *et al.*, 1993).

5.3.3 Some highlights of the scientific advances concerned with heterogeneous chemistry

Heterogeneous reactions are intrinsically more difficult to study than reactions in the gas phase as the compounds and intermediates are generally harder to characterise and the experiments themselves are more difficult to define and control. Nevertheless appreciable progress has been made within HALIPP, and the some parts of the subject are closer to the required quantitative understanding.

Table 5.5 lists the principal areas of study within the subproject, their present status and potential applications (Warneck *et al.*, 1994). Also section 3.3.2 lists a number of results that are of direct relevance to the formation of acidity in clouds.

Table 5.5:   Scientific advances within HALIPP

Topic	Status
Aqueous chemistry (Applications in the modelling of cloud-chemistry)	
Oxidation of S(IV): peroxo-compounds	H_2O_2, HSO_5^- completed, other reactants continuing
Oxidation of S(IV): ozone	Mechanism still uncertain
Oxidation of S(IV): catalysed by transition metals	Determination of mechanism for Fe(II)/Fe(III) catalysis completed; still uncertain for other metal ions or their biogenic complexes; ongoing activity for sulfonation of aromatics
Reactions of radicals and of oxygen with ions of transition metals	Ongoing activity
Termination reactions of S(IV) radicals	Significant progress, but ongoing activity
OH reactions	Completed; studies may have to be resumed when cloud species are better known
NO_3 reactions	Significant progress, but ongoing activity
Photochemical processes and quantum yields	Studies completed fort SO_3^{2-}, HSO_3^-, NO_3^-, NO_2^-, HNO_2^-, H_2O_2, $FeOH^{2+}$, $Fe(C_2O_4)^+$
Interaction of S(IV)-NO_x system; mixed S-N oxides and their hydrolysis	Ongoing activity

continued on next page

Table 5.5 continued

Topic	Status
Gas-liquid interactions	
(Applications in the modelling of cloud-chemistry)	
Mass accommodation coefficients:	Studies completed for SO_2, HNO_2, HCl, N_2O_5, NO_2, PAN, NH_3, HONO, HCOOH, CH_3COOH, CO_2, O_3; temperature effects and radicals remain to be studied
Henry's Law and hydrolysis coefficients	Alkyl nitrates, bifunctional nitrates, peroxynitrates, carbonyl compounds, acids. More information on cloud-water constituents required
Aerosol surface reactions and photocatalysis	
(Modelling of heterogeneous processes)	
Reactions of NO_2 with NaCl, sea salt, amorphous carbon	Nearly completed
Reaction of N_2O_5 with NaCl and sea salt	Nearly completed
HONO formation from NO_2 reacting with surfaces	No decisive progress
Photocatalytic activity of minerals	Degradation of organic compounds on TiO_2, Fe_2O_3, quartz, muscovite have been explored; additional studies are necessary
Lewis acid centres	Nearly completed
Sticking coefficients	α-pinene on quartz completed
Condensable products from gas-phase reactions	OH reaction with pinene is an ongoing study; exploration of other reactions needed

Warneck, 1995.

a. Oxidation of sulfur in clouds: elementary reaction steps

Clouds are thought to play an important role in the oxidation of sulfur dioxide (SO_2) to sulfuric acid, and hence in the acidification of precipitation. The SO_2 dissolves, together with the oxidants ozone (O_3), hydrogen peroxide (H_2O_2) and oxygen itself, in the cloud droplets where the conversion to sulfuric acid takes place in a complex sequence of reactions. It is likely that some of the reactions are catalysed by compounds of iron, copper and other metals that find their way into the cloud droplet from the aerosol particle on which the droplet forms by the condensation of water vapour. Using a variety of ingenious techniques (section 5.3.5) to identify the short lived ionic and free radical intermediates and to study their reactions in solution, many features of the complex catalysed reactions have been unravelled. It is expected that by the end of the project a definitive mechanism for sulfur oxidation in clouds will have been produced that can be used in the cloud modules of the simulation models sequences (Buxton *et al.*, 1992; Exner *et al.*, 1992; Lagrange *et al.*, 1993; Sedlak *et al.*, 1993; Zang *et al.*, 1993; Elias *et al.*, 1994, Botha *et al.*, 1994; Warneck *et al.*, 1994; Warneck and Ziajka, 1995).

b. Cloud chemistry: the detailed physical and chemical steps

In order to react within a cloud droplet, SO_2 and oxidant molecules in the gas phase must first diffuse to the droplet through the air and then pass through the surface into the droplet. Either the diffusion through the gas or the absorption on the surface can be the rate determining step, and it is essential to know the relative importance of these processes. One parameter describing them is the sticking, or accommodation, coefficient of the gas on the surface, and several elegant methods for determining these coefficients have been devised in order to provide a quantitative description of the absorption process. A noteworthy contribution has been the determination of accommodation coefficients for reactive gases that are difficult to handle (Kirchner *et al.*, 1990; Ponche *et al.*, 1993; Bongartz *et al.*, 1994; Hesse *et al.*, 1994; Bongartz *et al.*, 1995). Determinations have also been made of Henry's law coefficients that provide a measure of the equilibrium concentration of a gas in solution when the gas phase concentration is known (Kames *et al.*, 1992; Kames and Schurath, 1995).

c. Chlorine atom formation from aerosol surface reactions

A finding which possibly has a direct bearing on the formation of photo-oxidants in coastal areas is the observation that atomic chlorine can be formed from sea spray in the presence of nitrogen oxides, ozone and sunlight, through the formation and subsequent photolysis of the compound, $ClNO_2$. The process could contribute appreciably to the formation of photochemical oxidants in the early morning. Identification of a source of free atomic chlorine in the lower atmosphere must be given due consideration in models, and its impact properly assessed (Zetzsch *et al.*, 1992).

d. Surface processes: further topics of study

Other problems concerned with surface reactions of potential atmospheric importance that are being studied in HALIPP include (i) the stability of HNO_3 in aerosols or droplets, and the possibility of partial conversion to NO_2 after drying; (ii) aerosol formation from biogenic hydrocarbons; and (iii) formation of noxious compounds, such as phosgene or chlorinated hydrocarbons, by heterogeneous reactions. Photocatalysis on surfaces may also play a part in the chemical transformations of anthropogenic and biogenic species in the atmosphere (Grgic *et al.*, 1993; Guillard *et al.*, 1993; Tabor *et al.*, 1993; Vinckier *et al.*, 1993).

5.3.4   The incorporation of laboratory results into simulation models

An unsatisfactory aspect within the study of atmospheric chemistry has been the haphazard way in which new data from laboratory studies has been incorporated in the simulation models, used to understand the chemistry of the atmosphere. The chemical modules within the models, while in principle containing the same scheme of chemical reactions, actually differ appreciably from each other, since each is a simplification of the natural situation and there is no agreed way to affect such simplifications. Thus both the initial formulation of the schemes of chemical and cloud reactions, used in the models, and their updating depends on the skill, whim and enthusiasm of the individual modellers who construct and maintain them. In an attempt to improve the process, two inter-subproject groups were formed within EUROTRAC: the Chemical Mechanism Working Group (CMWG) and the Cloud Group (CG) each of which consisted of invited investigators from the appropriate subprojects.

a. Chemical Mechanism Working Group (CMWG)

The activities of the CMWG, in which investigators from LACTOZ, TOR and EUMAC, together with invited investigators from EMEP (section 2.3.1) participated, are shown in Table 5.6.

Table 5.6:   Scientific advances of the CMWG.

Topic	Status	Application
First scrutiny of EMEP chemical mechanism	Finished	Policy making
Improvement of aromatic and biogenic oxidation in EMEP mechanism	In progress	Policy making
Intercomparison of chemical schemes	In progress	Policy making
Comparison of field data and model calculations: fast photochemistry of OH	In progress	Validation of chemical schemes
Comparison of field data and model calculations: photo-oxidants (O_3, PAN, HCHO, CH_3CHO, ROOH)	In progress	Validation of chemical schemes

Poppe, 1995

An important early result was the scrutiny of the chemical module of the EMEP photo-oxidant model, that was carried out by a group from LACTOZ (Wirtz *et al.*, 1994). The details of the model, which is used for policy studies, were largely confirmed but also some useful suggestions for improvement were also provided. Most of the changes recommended have already been incorporated into the EMEP model and the next step envisaged is for the LACTOZ group to provide better schemes for the oxidation of aromatic and biogenic compounds (see section 5.3.2.b).

b. The Cloud Group (CG)

The CG was formed from HALIPP, GCE, ALPTRAC, GLOMAC, EUMAC investigators. The aim was an exchange of information between investigators in laboratory and field studies and modellers in order to develop a better understanding of cloud processes. Appreciable progress has been made with the utilisation of HALIPP reaction schemes in some cloud models and the publication of a reaction scheme for cloud chemistry (Floßmann and Möller, 1995).

5.3.5 Developments of laboratory techniques within the programmes

As indicated in section 5.2.1, progress in science is usually accompanied by improvements to experimental techniques and instrumentation or by devising new methods and techniques. As Tables 5.7 and 5.8 indicate, state of the art instrumentation was employed in the laboratory work and, in many cases, new or improved techniques permitted experiments to be performed that were hitherto difficult or even impossible. Much of its progress in the heterogeneous work is due to the ingenuity of its investigators who have invented, adapted and developed techniques to enable them to study the processes at a quantitative level.

Table 5.7:   Developments of instruments and techniques within LACTOZ.

Technique	Application
Methods for kinetic studies	
Environmental chambers and static photoreactors Volume: 10–1000 litres Temperature control (–30 to + 50 °C)	The reaction can be followed after filling the reactor with the reactants. There is a quasi-steady state production of radicals. The system is used in conjunction with relative rate measurements and applied for product studies. Can be operated at atmospheric pressure and ambient temperature.
Discharge flow	Production of radicals by a discharge with flowing reactants and a carrier gas with constant flow through a cylindrical tube. Reaction time is measured along the flow (1 ms to 1 s time scale). Operated at reduced pressures and variable temperatures.
Discharge stopped flow	After mixing of the reactants, the flow is stopped, and the reaction can be followed as a function of time (100 ms to 100 s time scale)

continued on next page

Table 5.7 continued

Technique	Application
Flash photolysis	A pulsed light flash is used to generate radicals photochemically within a short period. After the light pulse, the reaction can be followed in real time (100 ns to 100 ms time scale). Pressures of well in excess of 1 atmosphere can be used.
Laser photolysis	A variant of the flash photolysis technique using a laser as the light source. Highly monochromatic radiation can be used. Times down to picoseconds or even femtoseconds can be probed in special cases.
Modulated photolysis	A light source is repeatedly turned on and turned off, and the variation of concentrations of reactants, products and intermediates (*e.g.* radicals) is probed, usually spectroscopically. Time scales are typically of the order of 0.1 s to 100 s. Pressures of up to one atmosphere may be used.
Pulse radiolysis	Pulsed beams of high-energy electrons are generated. These electrons produce, after a complex sequence of reactions, radicals whose reactions can be followed in real time

Techniques used to determine concentrations of reactants, products, and reactive intermediates

Fourier-transform IR spectroscopy (FTIR)	High-sensitivity, high-resolution IR absorption spectroscopy.
	Can be used with long-path (multiple pass) cells for enhanced sensitivity(pathlengths 50 to 1000 m). Spectral resolution $0.1–1 \text{ cm}^{-1}$.
	Sensitivity allows time for complete scans of *ca.* 1 min.
Tunable diode laser (TDL)	Ultra high resolution technique (largely IR)
	Can be used with long-path cells for enhanced sensitivity
	(pathlengths 50 to 1000 m). Spectral resolution 10^{-4} cm^{-1}.
	Sensitivity (relative absorption down to 10^{-6} with modulation) allows time resolution for restricted scan of *ca.* 1 s.
UV or visible absorption (UV/visible)	Conventional absorption spectroscopy using spectrometers.
	Long path lengths can be used.
	Improvements in sensitivity and/or resolution with diode arrays.

continued on next page

Table 5.7 continued

Technique	Application
Resonance fluorescence (RF)	Excitation of fluorescence by spectral resonance emission line.
	Highly sensitive and species specific. Good for atoms, OH, *etc.*
Laser induced fluorescence (LIF)	Excitation of fluorescence by narrow laser emission line (usually tunable). Highly sensitive and species specific. Good for OH, NO, NO_2, NO_3, CH_3O, *etc.*
Laser long-path absorption (LLA)	Laser beam (usually CW) used for long pathlength absorption.
Gas chromatography (GC) High-pressure liquid chromatography (HPLC)	Separation of species by a chromatographic column. Detection can use a variety of methods (flame ionisation, electron capture, UV absorption, *etc.*)
Mass spectrometry (MS)	Separation of species by molecular mass. The species are first ionised, and the masses resolved by electric or magnetic fields.
Gas chromatography – Mass spectrometry (GC-MS)	Combination of GC to separate species by retention time on a column and detection by MS for mass resolution

Becker 1995

Table 5.8:   Developments of instruments and techniques within HALIPP.

Technique and Purpose	Description and Application/Advances
Stopped-flow reactor Determination of the mechanisms and reaction rate coefficients for elementary reactions.	Reactant solutions are rapidly mixed and forced into a flow tube. Spectra of reactants and/or products are taken in real time after the flow is suddenly stopped.
	Improvements to commercial instruments have been made, and use of fast UV/visible scan spectrometers has been exploited within HALIPP.
Pulse radiolysis. Determination of the mechanisms and reaction rate coefficients for elementary reactions.	Ionisation of water by fast electrons, conversion of primary to secondary reactants by addition of suitable solutes, spectro-optical detection in real time of labile intermediates and/or stable products.
	Development of a fast premixing device for use with pulse radiolysis facilitated study of unstable S(IV) compounds.

continued on next page

Table 5.8 continued

Technique and Purpose	Description and Application/Advances
Laser flash photolysis. Determination of the mechanisms and reaction rate coefficients for elementary reactions.	Generation of radicals by flash photolysis of suitable solutes in the aqueous phase, conversion of primary to secondary species, if necessary, by addition of suitable reagent, spectro-optical detection of intermediates and/or products. Construction of a multi-path optical cell of improved sensitivity.
Steady state photolysis. Determination of quantum yields for photochemical processes. Generation of radicals for mechanistic studies.	Aqueous solution is irradiated with UV or visible light. Products from radical scavengers are analysed. Although conventional, the technique has been extended to use with scavengers for OH and SO_4^- radicals and has led to important new results.
Continuous gamma-radiolysis. Determination of chain length for propagation reactions in systems with known termination reactions. To study the role of HO_2/O_2^- in Fe and Cu redox cycles.	Irradiation of an aqueous solution with gamma-rays from a ^{60}Co source, analysis for products as function of duration, intensity, *etc*. The only method so far that has provided a reliable value for the rate coefficient of the propagation reaction in the Bäckström mechanism of sulfur(IV) oxidation. Simulation of HO_2/O_2^- reactions in clouds.
Radical scavengers. Test of mechanisms and determination of reaction rate coefficients by competition with reaction having a known rate coefficient.	The addition to reacting solution of one or more reagents, which interact with radicals causing a change in experimental conditions. Widely used conventional technique, which relies on ingenuity of experimenter to give unequivocal results.
Conventional study of aqueous reactions. Mainly for exploratory study of reaction behaviour, determination of products.	Follow decay of reactant and build-up of products using suitable analytical techniques. Advances lie in the exploitation of new analytical techniques.
Falling-droplet train. To study the transfer of substances from the gaseous to the aqueous phase and to determine accommodation coefficients.	A vibrating orifice is used to generate a train of water drops about 50 µm in size. After precipitation through a gas column, the drops are captured and analysed for materials incorporated from the gas phase. Major technological development in Europe: only two such pieces of apparatus exist in the world.

continued on next page

Table 5.8 continued

Technique and Purpose	Description and Application/Advances
Liquid-water jet. To study the transfer of substances from the gaseous to the aqueous phase and to determine accommodation coefficients.	A cylindrical free-liquid water jet of about 0.1 mm diameter is brought into contact with a gas layer before the water enters a receiving orifice for collection and analysis. This technique was newly developed for the HALIPP subproject, and is a major breakthrough. It provides a flexible alternative to the falling-droplet train technique.
Gas-bubble water column. To determine Henry's law coefficients for compounds that react with water and decompose.	A carrier gas containing a soluble substance is bubbled through a vertical water column; effluent gas and water are analysed. Henry's law and reaction coefficients for PAN-like compounds and organic chlorine compounds have been determined; this is an improved method.
Two-phase chemical reactor. To study the uptake and oxidation of gaseous components by and in aqueous droplets.	Droplets with a narrow size distribution are generated in a nebulizer from an aqueous salt solution and admixed with a carrier gas that is made to flow through a reactor. The oxidation of sulfur dioxide in aqueous droplets has been explored in this system; its advantage is a realistic simulation of cloud chemistry.
Floating drop. To study the interaction of gaseous components as well as aerosol particles with raindrop-sized water drops in the laboratory.	Uptake of an acidic gaseous component by a water drop of size ca. 1 mm floating in an upward gas flow is determined optically by the colour change associated with an acidity indicator. The apparatus was constructed and tested with sulfur dioxide as reagent; the necessary theory was developed.
Wetted-wall flow tube. To study surface interactions of gaseous compounds with water and solutes in the aqueous phase.	The interior surface of a glass tube, wetted with an aqueous solution, is exposed to a gaseous reactant; effluent solution and gases are analysed for reaction products. The method was newly developed in order to determine products resulting from the interaction of nitrogen pentoxide with sodium chloride and sea salt.
Coated-wall flow tube. Determination of gas-solid sticking coefficients.	A gaseous substance is made to flow through a glass tube internally coated with the material to be studied. Losses to the wall are measured by analysis of the effluent. The method has been tested by low pressure mass spectrometry to see if it can be used to determine the sticking coefficient of α-pinene on various surfaces.

continued on next page

Table 5.8 continued

Technique and Purpose	Description and Application/Advances
Fluid-bed reactor Study of heterogeneous reactions and the role of water in such reactions	Aerosol-like materials contained in a reactor bed are exposed to suspected reactants; losses by adsorption and reaction products are determined. To explore the catalytic activity of iron oxides, titanium dioxide and silicates as surrogate aerosol materials in the heterogeneous oxidation of certain organic compounds.
Knudsen cell surface reactor. To study heterogeneous chemical processes in detail.	Low-pressure device allows *in-situ* mass spectrometric analysis of effluent gases resulting from surface reactions. Used to explore the interaction of nitrogen dioxide with various surfaces.
Calorimetry of active sites. Determination of the number of Lewis acid centres created.	A solid sample is irradiated by UV light or treated thermally to produce active sites followed by reaction with water or another reagent; calorimetry is used to determine the heat evolved. Developed to study tropospheric aerosol constituents such as fly ash, aluminium oxides, desert sand, common soils.
Environmental aerosol chamber. To study the interaction of radicals and gases with the aerosol surface; to study the degradation of organic materials adsorbed on the aerosol surface.	A chamber of *ca.* 1 m^3 capacity is needed to keep aerosol particles suspended for some time; external lamps allow irradiation with simulated sunlight. Determination of accommodation coefficients for interaction of nitrogen pentoxide with sodium chloride and sea salt particles; exploration of effect of muscovite, titanium dioxide and fly ashes on the photodegradation of organic compounds.
Gas-liquid-solid reactors. Testing mechanisms and rates at very low concentrations of otherwise unobtainable reactants	Supplying the reactants from gas and solid phases to stabilise the concentration of dissolved S(IV) at limits determined by mass transfer and solubility product. Unconventional use of chemical engineering techniques to explore reaction behaviour at limiting concentrations.

Warneck, 1995

5.3.6   Needs for the future

Particular gaps in our present knowledge that will have to be filled by laboratory work are:

in the gas-phase chemistry of photo-oxidants:

- elucidation of the detailed mechanisms for the oxidation of aromatic compounds;

- elucidation of the detailed mechanisms for the oxidation of biogenic VOCs such as isoprene and the more common terpenes;

- determination of the reactions of alkoxy radicals produced in the oxidation of higher hydrocarbons;

in heterogeneous chemistry:

- testing and validation of chemical mechanisms for oxidation in cloud droplets;

- study of concentration effects on the solution reactions that occur in cloud droplets;

- study of possible atmospheric reactions on aerosol surfaces (the development of new techniques and methods will certainly be required);

in chemical mechanisms:

- scrutiny and comparison of the lumping schemes used to simplify many chemical models;

- possible use of automatic, mathematically-steered methods for the simplification of chemical mechanisms;

- incorporation of the heterogeneous mechanisms into more realistic cloud models.

5.4   Models

5.4.1   Introduction

Simulation models, which relate emissions to atmospheric concentrations or deposition, are powerful tools in policy-oriented applications. In these types of models the contribution of specific emission sources to the occurring concentration or deposition can be determined, and the result of abatement strategies can be analysed. However, this can only be done with confidence when these models represent reality in the correct way.

Simulation models, because they contain modules or sub-models describing processes, provide a way to integrate scientific knowledge and make it available for policy development.

Finally, simulation models are also scientific tools. By performing sensitivity calculations, the relative importance of different atmospheric processes can be investigated. New questions are raised, and there is a feedback to field measurements, instrument development and laboratory studies.

Before presenting an overview of the models developed within EUROTRAC, several general points about the models should be made.

There are two types of simulation models: Lagrangian and Eulerian models. In the Lagrangian model, an air parcel of a specific size is followed along a trajectory until it reaches the receptor point. The assumption made is that the air parcel retains its identity along the trajectory. The advantage of trajectory models is their relative simplicity, so that a large number of model runs can be performed in a short time. They are especially suited to cases where the influence of emissions on a specific receptor point is to be considered. However, to make calculations over a large area, the whole area has to be covered with receptor points.

An Eulerian model covers the area under consideration with a grid of a specific size, and it has a specific number of vertical layers. The continuity equation is solved for the whole array of boxes simultaneously. The advantage of Eulerian grid models is that, with one model run, the whole area is covered; also they are capable of considering the complete meteorological situation both vertically and horizontally. Both Eulerian grid models and Lagrangian trajectory models can be used to study photo-oxidants or acidification, and can be directed towards long-term, monthly to yearly averages, or short term episodes of a few days to several weeks situations. In EUROTRAC the focus of the research was on Eulerian grid models.

In the following sections an overview will be given of the major simulation models developed and investigated in EUROTRAC, on continental, urban and global scales.

5.4.2   Regional simulation models

a. Early models

Before the start of EUROTRAC, several regional simulation models already existed which covered the whole of Europe.

The EMEP trajectory model: For studying acidification or nutrification EMEP uses a one-layer trajectory model that has been used for the last 20 years, with continuous improvement, to calculate long-term averaged concentrations and country budgets of SO_x, NO_x and NH_3 over Europe. For photo-oxidants, the EMEP trajectory model can also incorporate a photochemical gas phase module, and has applied it to photochemical episodes (Eliassen, 1982). It is used to support policy decisions in the framework of UN-ECE (Eliassen, 1975).

The PHOXA model: An Eulerian grid model in the PHOXA program had been developed by Klug also to calculate acidification and nutrification over Europe (Klug, 1989). The Eulerian grid model RTM-III was adopted, modified and applied to determine photo-oxidant formation during meteorological episodes (Stern and Builtjes, 1989).

In the framework of the PHOXA program, the TADAP model was developed which addressed both acidification or nutrification and photo-oxidants. It covers the whole troposphere and includes a fairly detailed description of clouds (Stern and Scherer, 1989).

All these models use actual meteorological information as input: thus they make use of interpolated and analysed meteorological data, in one to six-hourly intervals, based on actual meteorological situations for specific days, months or years. The difference between the models lies in the treatment of the dispersion processes, in the Lagrangian or an Eulerian framework, and in their vertical structure. In principle, the finer vertical structure requires an Eulerian framework so that the more meteorologically coherent data can be used, which implies more reliable results. However, the reliability also depends strongly on the accuracy of other parts and inputs of the model, such as the emissions and the chemical scheme used.

b. EUROTRAC models

The EURAD model: To investigate the functional relationship between emissions and concentrations or deposition, the Eulerian grid model developed in the US at NCAR by J. Chang has been adopted in EUROTRAC, modified and applied in the EUMAC subproject (Chang *et al.*, 1987, Stockwell *et al.*, 1990).

The central model of EUMAC is EURAD (European Acid Deposition Model) that is intended for the simulation of acidification, nutrification and photo-oxidants over Europe under specific episodic conditions of up to about two weeks. The model contains the RADM-2 gas-phase mechanism and an aqueous-phase scheme (RADM-2, Hass, 1991). The model encompasses the whole troposphere and the lower stratosphere, and covers all of Europe in a grid of 80×80 km^2 or less. It can be used for nesting with a resolution of less than 3×3 km^2 (Jacobs *et al.*, 1995). The model is connected to a sophisticated meteorological model, currently MM4/MM5 (Meteorological Model 4/5), which uses analysed meteorological data from ECMWF (The European Centre for Medium Range Weather Forecasting) to calculate the required meteorological input data. The EURAD model system is complemented by the EURAD emission model (Memmesheimer *et al.*, 1991) using data from EMEP and GENEMIS as input. The model can be run in a forecast mode for operational applications in air pollution prediction.

One important question, both scientifically and for abatement policy, is the question of the phenomena which determine the boundary layer and the tropospheric budgets of trace gases, such as ozone. A determining parameter in such budgets is the vertical fluxes of the trace gases. In meteorology and weather forecasting, vertical fluxes and parameters such as the mixing height, are of secondary importance; however for trace gas budgets, they are of primary importance. Consequently, EURAD has been used to study vertical exchange, the results being compared with measured vertical profiles. The work will lead to an improved description of vertical fluxes of water vapour, for example, in meteorological models (Ebel *et al.*, 1991, 1993).

It is worth saying that simulation models such as EURAD are complicated tools which require experience to handle them. Consequently specific methods have been developed to perform data analysis and diagnosis on the enormous amount of output produced by a single model run.

Although EURAD has been used mainly for scientific studies, it is intended for use in comprehensive studies of atmospheric processes affecting emission reduction measures, and for the design of air pollution abatement strategies. Assessment of emission reduction strategies, the determination of source-receptor relationships and transboundary transport is possible using EURAD. In particular, abatement strategies for tropospheric ozone, which require the determination of the effects of VOC or NO_x controls, can be investigated in detail. Also, in combination with global models, abatement strategies for tropospheric ozone as a greenhouse gas can be evaluated. The model is used for assessing the impact of stratospheric ozone on the troposphere through the simulation of cross tropopause ozone fluxes. It has been applied to the old question of the impact of air traffic on tropospheric ozone using a special chemical mechanism for tropopause conditions (Perry *et al.*, 1994). Furthermore the validation of emission data can be performed by comparing results with the measured concentrations, as well as the design of optimal observation networks (see Volume 7 in this series).

The HIRLAM/EURAD-CTM system: The HIRLAM/EURAD-CTM system has applications in the area of regional problems in global modelling as carried out in GLOMAC. The Chemical Transport Model (CTM) of the EURAD system has been combined with the High Resolution Model (HIRLAM) nested into ECHAM (the climate model of the Max-Planck Institute for Meteorology at Hamburg; Langmann, 1995). This activity of EUMAC underlines the necessity of integrating global and regional modelling to the assessment of future climate and air pollution development.

The FINOX model: The Finnish limited area model for oxidised nitrogen compounds (FINOX) has been used in EUMAC to study the role of large areas of water for photo-oxidant formation in northern Europe. In particular, problems related to the Baltic Sea have addressed. FINOX is an Eulerian grid model with about 50 km or higher resolution (Hongisto, 1992).

The LOTOS model: In TOR the LOTOS (Long term ozone simulation) model, an Eulerian grid model, has been improved and applied. LOTOS is intended for the simulation of photo-oxidants over Europe for specific episodes and for longer time periods of a growing season and a year. It can be regarded as a successor to the episodic RTM-III model.

The LOTOS model covers the lower part of the troposphere, up to about 3 km, and covers all of Europe in grid squares of $0.5° × 1°$, which is about $60 × 60$ km^2. The model uses the CBM-IV chemical scheme and requires detailed meteorological input. The LOTOS model is simpler than models like EURAD, and longer time period calculations are possible; so in this way it complements EURAD (Builtjes, 1992).

The LOTOS model development has been used to study key questions in TOR, which concerns the study of the tropospheric ozone budget, see Volume 6 of this series.

The LOTOS model has also frequently been used to answer policy related questions such as the effectiveness of proposed abatement strategies, the influence of biogenic VOC emissions on abatement strategies, and the calculation of the so-called POCP (Photochemical ozone creation potential) values for different VOCs.

c. Other models

A continental chemical transport model coupled to a numerical weather prediction model, developed at the University of Bergen in TOR, has been used to study the chemical composition of the troposphere over Europe, as well as over the North Atlantic and North America. Recently, it has been applied to study ozone formation from subsonic aircraft and the impact of stratospheric intrusions (see Volume 6 in this series).

It should be mentioned that in TOR several Lagrangian trajectory models have been applied to analyse specifically the measurements performed at the TOR stations.

Within ASE, a comprehensive trajectory, atmospheric chemistry and deposition model (ACDEP) has been developed and used to estimate the nitrogen deposition in Danish waters. It uses a nested grid ($25 × 25$ km^2) for Denmark and $150 × 150$ km^2 for the rest of Europe. (Larsen *et al.*, 1993)

d. Model intercomparisons

Of special interest is an intercomparison study in which two EUROTRAC models, EUMAC and LOTOS are compared in their performance and behaviour for a specific period in July and August 1990 with two non-EUROTRAC models, the EMEP model and REM-III from the Free University of Berlin.

One of the aims of this intercomparison study was to investigate the possible improvement of model performance as a result of a more refined description of important processes contained in the models. The final report of this study will be available at the beginning of 1996.

An overview of EUROTRAC simulation models is given in Table 5.9.

5.4.3   Urban simulation models

Most people in Europe live in urban areas, and in most urban areas there are considerable exceedances of air quality standards.

Consequently, the knowledge concerning the relationship between local emissions and local concentrations and the influence of sources outside the urban areas on the air quality in these areas is of high political relevance. Thus phenomena on a smaller, urban scale are of importance, in areas of about 50×50 to 200×200 km^2.

Considering photo-oxidant formation on a local scale, before the start of EUROTRAC, the so-called UAM (urban airshed model), an episodic Eulerian grid photo-chemical model, was used in Europe to study photochemical oxidant formation in the Rijnmond area of the Netherlands (Builtjes and Reynolds, 1983), and in the Köln-Bonn area in Germany. More recently it has also been used by the University of Aveiro for the greater Lisbon area, as an integrated model system (MAR) which couples the CSU meso-meteorological model with UAM.

In the EUMAC project, considerable attention has also been given to local scale phenomena, since the concentrations on a local scale are often influenced substantially by pollutants imported from outside the local area. Using the EURAD model, two approaches are possible.

First, the EURAD model can calculate the boundary conditions for the local area. These boundary conditions are subsequently used by a local model for calculations on a smaller grid. This is called a one-way nesting or zooming model. The local model uses the boundary conditions of the larger scale model.

The second possibility is two way nesting: inside the EURAD model a finer grid is used for the local area under consideration which is completely embedded in the larger scale model.

In the EUMAC project, smaller scale urban air pollution problems have principally been addressed by developing the EUMAC Zooming Model (EZM), which is an Eulerian one-way nested grid model within EURAD. EZM addresses both the meteorological and photochemical phenomena on an urban scale, with grids of about 5×5 km^2 (Moussiopoulos, 1994). Nesting thus replaces the meteorology using MM4/MM5 (MEMO/MARS) and the photochemical dispersion model.

A number of application studies have been performed using EZM, especially focusing on pollution problems of the Mediterranean like Athens and Barcelona, but also other areas like Heilbron, Graz and the Black triangle (see Chapter 2). All these studies for particular areas generally have a local policy question as their basis. In Athens the abatement of the high ozone levels during the summer is the issue; in Barcelona the issue was the photochemical oxidant formation during the 1992 Olympics and its abatement; in Heilbron it is the investigation of the effectiveness of local abatement measures on precursor concentrations and ozone formation.

A comparison was made in the greater Lisbon area between the EZM and the MAR system for a situation where the coastal breezes are quite strong (Borrego *et al.*, 1995). The study indicated how certain features of the wind fields generate areas in which the ozone concentrations exceed the legislative guidelines.

An extensive model intercomparison study, APSIS, has been performed for the Athens area, which has revealed many phenomena peculiar to Athens (Moussiopoulos, 1993).

The EZM could be used in the future for assessing measures to limit primary pollutant emissions as well as to forecast air pollution episodes on a local scale. It should however be noted that the usefulness of local scale calculations is often restricted due to lack of the required information on detailed local emissions on a fine grid resolution.

Another effort within EUMAC is the further development of the CIT transport model and its application to ozone control and abatement strategies over urban and suburban areas (Giovannoni, 1993). The model has already been tested within the APSIS simulations in the Athens region, and has been used as well for the Swiss Plateau ozone numerical simulation (SPONS), and the Zürich ozone numerical simulation (ZONS), with grid sizes ranging from 1×1 km^2 to 5×5 km^2. The chemical module used for the simulations is the Lurman, Carter, Coyner (LCC) mechanism coupled to a condensed-phase reaction mechanism.

In EUMAC and also in the EUROTRAC project TRACT, meteorological and chemical calculations on a local scale of 300×300 km^2 have been performed with the same chemical code as that used in EURAD, using the KAMM/DRAIS model. Thereby nesting procedures have been developed for combining the larger scale EURAD with the smaller scale KAMM/DRAIS model system (Nester *et al.*, 1995).

In the ASE subproject a simulation model, similar to the EMEP trajectory model, has been developed to address the deposition of NH$_3$ to sea areas with a resolution of 30×30 km^2 for a 400×400 km^2 area. This model too is a valuable tool which can give answers to policy related questions (Larsen *et al.*, 1994).

5.4.4   Simulation models on a global scale

Simulation models were first developed to address urban scales, followed by regional and continental scale models, but about 1980, simulation models were also developed to cover the entire world.

The first global models were 2-D global models, which averaged the global concentration in an east-west direction using a mixing period of about 2 weeks. Crutzen (1983) and Isaksen (1987) first used 2-D global models to address the tropospheric budgets of CO, CH_4 and ozone. Climatological data were used to describe the meteorology. This type of model is still in use today, because it is capable of handling large chemical schemes, and can be considered as reliable, especially for trace gases with an atmospheric residence time of more than about one month.

The first 3-D global Eulerian grid models were developed in the early 1980s, most notably the Moguntia model in Europe. It was developed by Zimmerman (1987), and still uses climatological input like the 2-D global models. At the outset, the chemistry of Moguntia was very limited.

The GLOMAC subproject attempted two major areas of work. The first was to improve the Moguntia model, and the second was the development of a global simulation model, similar to regional models, using actual meteorology rather than climatology.

The Moguntia model has been improved, and currently includes gas-phase and aqueous-phase chemistry and a climatological input (Dentener and Crutzen, 1993).

A first step in the direction of a global model using actual meteorology is TM-2. This model uses ECMWF meteorology in an off-line mode and includes some chemical reactions. A further global model, ECHAM, is under development; it uses ECMWF meteorology directly together with a limited chemical scheme.

Each of these 3-D global GLOMAC models necessarily has a rather coarse resolution, from about $3° \times 3°$ up to $10° \times 10°$. Therefore, a nesting procedure using HIRLAM with the EURAD-CTM (section 5.4.2) has been developed.

The GLOMAC models are primarily meant to answer basic scientific questions concerning the budgets of trace gases on a global scale, most notably the tropospheric ozone budget.

Investigations have also been performed on the influence of the NO_x emissions from lightning, the influence of sources from forest fires in the boreal zone and the influence of biogenic emissions from vegetation. A key parameter of the research is the OH radical, because of its fundamental importance for the oxidising capacity of the atmosphere.

Some policy-related applications are as follows. The Moguntia model has been used to calculate future trends of CH_4 (IPCC framework), CO and O_3, and has indicated the anthropogenic contribution to the increase of tropospheric ozone (Lelieveld, 1993). Moguntia has also been used to address the chemistry of HCFCs (Kanakidou, 1993) and the influence of sulfate aerosols on the radiation budget of the troposphere (Raes, 1993). Recently the influence of aircraft emissions especially on the ozone concentrations in the upper troposphere and the lower stratosphere has been addressed by Moguntia. First attempts have also been made to apply TM-2 to the same question.

The Moguntia model is a tool which can address policy related questions on a global scale, as ECHAM will be able to do in the future.

Table 5.9 gives an overview of the EUROTRAC simulation models.

5.4.5   The validation of simulation models

When simulation models are used to study scientific issues, or are used to provide policy-oriented answers, it is essential to assess their capabilities. Next to the question of what models can actually do, comes the question how well they can do it, and with what accuracy.

The simulation models described in the preceding sections all give the functional relationship between concentration or deposition and precursor emissions. Simulation models simulate numerically the physical, meteorological, and chemical phenomena which determine the spatial and temporal patterns of concentrations or deposition as a function of emissions. Because it is not possible to calculate these patterns at every point in time and space, due to our incomplete knowledge of input data and the restrictions in computer capacity, the results from these models are box-averaged concentrations or depositions.

The question as to what extent simulation models are capable of simulating true concentrations, and with what accuracy, cannot be answered in a straightforward manner.

Simulation models contain several sub-models or modules such as the chemical mechanism, the photolysis scheme, the dry deposition model, the parameterisation of the turbulent eddy viscosity and so on. All these sub-models are process-oriented models based on measurements, with their own inherent level of uncertainty. In addition, there are uncertainties in the input data. Meteorological input data are based on measurements and objective analysis together with a meteorological model. The meteorological simulation model, which contains empirical sub-models, has its own inherent inaccuracy. Last but not least the emission data required as input are associated with large uncertainties.

So the uncertainty of the calculated concentrations or depositions from a simulation model is influenced by the uncertainty of the input data, and the uncertainties associated with the descriptions of the processes involved.

Table 5.9:   Simulation models in EUROTRAC.

Acronym	Full name	Model type	Purpose	Applications
EURAD (subproject EUMAC)	European Acid Deposition Model	Eulerian grid grids 80×80 km^2 nesting down to 3×3 km^2. Europe 16 km height; 30 km for aircraft emissions	Episodic calculations of photo-oxidants and acid deposition	Evaluation of abatement strategies for episodes, influence of subsonic aircraft
LOTOS (subproject TOR)	Long-term ozone simulation model	Eulerian grid grids 60×60 km^2 Europe 3 km height	Episodic and long-term calculations of photo-oxidants	Evaluation of abatement strategies for episodes and long-term averages
(subproject TOR)	Chemical transport model University Bergen	Eulerian grid grids 80×80 km^2 part of northern hemisphere 15 km height	Episodic calculations of photo-oxidants	Evaluation of abatement strategies for episodes, influence of subsonic aircraft
EZM (subproject EUMAC)	EUMAC Zooming Model	Eulerian grid 5×5 km^2 local scale 3 km height	Episodic calculations of photo-oxidants	Evaluation of local abatement strategies
CIT (subproject EUMAC	Carnegie-Mellon Institute of Technology multi-phase transport model	Eulerian grid 1×1 km^2 to 5×5 km^2 2 to 3 km height	Episodic calculations of photo-oxidants in presence of clouds	Evaluation of local abatement strategies
Moguntia (subproject GLOMAC)	Model of the general universal tracer transport in the atmosphere	Eulerian grid $10° \times 10°$ Global 16 km height climatological input	Global tropospheric budget studies	Evaluation of global abatement strategies, influence of subsonic aircraft
TM-2 (subproject GLOMAC)	Transport model-2	Eulerian grid $4° \times 5°$ Global 20 km height off line meteorological input	Global tropospheric transport studies	Evaluation of global abatement strategies
ECHAM (subproject GLOMAC)	European Centre Hamburg Model	Eulerian grid $2.5° \times 2.5°$ Global 35 km height on line meteorological input	Global tropospheric and stratospheric budget studies	Climate response studies

The test of the quality of a simulation model itself should be a comparison between measured and calculated concentrations or deposition of a trace gas such as ozone for the time period under consideration. Then another problem arises; as already mentioned the model produces box and volume averaged concentrations, while measurements are made at a points. Only measurements which are representative of a whole box should in fact be used for such a comparison.

To illustrate the point, one can consider the calculation of hourly averaged ozone concentrations, which perhaps deviate by 30 % from the measurement of ozone on a specific afternoon at a specific time. In fact, 30 % is the value used by the US-EPA as the borderline between a good and a bad model. The 30 % difference could easily be caused by an error in the VOC emissions, and especially in the VOC speciation. Also a slight error in wind direction could mean that the calculated ozone plume originating from a precursor emission area downwind misses the actual measuring point which is in reality hit by the plume.

The question of which parameterised process, or of which input data set including their inherent inaccuracy is most important in determining the model output depends on the question addressed with the simulation model. It might well be that for the calculation of yearly averaged NH_3 concentrations on a continental scale, the dry deposition velocity of NH_3 is the most important parameter; for a similar question concerning NO_x it might be the mixing height, or even the stack height.

The calculations of episodic ozone are sensitive to the speciation of anthropogenic VOC emissions and, as has been shown recently, they seem to be less sensitive to which chemical mechanism is used (Stern, 1993). However, both statements assume that the total amount of anthropogenic VOC emissions is not in error by more than a factor of 2 or 3, and that the three chemical mechanisms considered do not have a common fundamental error.

There is another fundamental problem: because of the uncertainties involved in the process descriptions as well as the input data, and because of the non-linearities involved in the relationships between emissions and concentrations or deposition, it is never possible to prove that when a calculated concentration agrees with a measurement the model description is actually correct. The agreement could be due to a cancellation of errors (Oreskes *et al.*, 1994).

A particular problem arises when arbitrary correction factors to improve the agreement between measured and calculated concentrations are used. Although the agreement can be improved in this way for the particular conditions, it might be due to accidental cancellation of errors, and the reliability of such a model with respect to policy-oriented questions will be limited.

One final remark: because simulation models are often used to evaluate emission reductions, it is important to try to ensure that the models also calculate adequately the changes in, say, ozone for a given change in precursor emission. As it is not feasible actually to make the emission reductions and measure the resulting ozone pattern, except perhaps on a local scale as in the Heilbronn experiment, only careful model diagnosis can help to improve the confidence in the model results. Multi-component validation, and the investigation of spatial and temporal gradients in calculated and measured ozone can be used here.

Thus when a simulation model is used to evaluate the often very costly abatement strategies, the outcome of the model should be as accurate as possible.

One way to address this problem might be the following. When a simulation model uses the best available knowledge for its sub-models with state-of-the-art meteorological information and accurate emissions, all with a known accuracy and sophistication, the model should be considered as reliable in the sense that the outcome of the model is the best answer possible. It is important to stress again the point that these simulation models do not and should not use tuning parameters to improve their performance.

Furthermore, if the simulation model is tested as far as possible against measurements, and describes phenomena which also happen in reality, then this is a partial proof that the model should be considered as the best available tool. A model should show a one to one relationship between measurements and observations, and, to the best of our knowledge, be reliable. Obviously, new proven scientific results have to be continuously incorporated in the models to keep them up to date.

From these remarks it will be clear that the quality of the models deserves further extensive attention: for example, a weak point in current simulation models is that they do not provide a calculated value with a given level of uncertainty, and that, at the moment, no methodology exists to do this consistently.

5.4.6 Process-oriented models

Process-oriented models cannot be applied directly as they do not give the relationship between emissions and concentrations or deposition, but they are used in the form of sub-models as a part of simulation models.

It should be stressed that the development of these process-oriented models is essential for the increase in scientific understanding, and the improvement of simulation models. These should be viewed as valuable tools from EUROTRAC which as sub-models have and will be applied in applications oriented to policy.

In this section, three types of relevant process-oriented models will be addressed. Attention will be given to chemical mechanisms, cloud models and models describing the photolysis process.

a. Chemical mechanisms

Laboratory studies, as described in detail in section 5.3, focus on investigations to reveal the chemical transformations occurring in the atmosphere, and lead to quantitative information concerning reaction rates, reaction products and photolysis rates.

A full chemical gas-phase or liquid-phase mechanism contains hundreds of species and thousands of reactions. Incorporation of such a chemical scheme into a simulation model is far beyond the capabilities of current computers, and would also be partly out of balance with the current knowledge and accuracy concerning emission data. Consequently condensed schemes are used in simulation models.

The chemical mechanism working group (CMWG) was established in EUROTRAC to form a bridge between laboratory studies and simulation modelling. Under its auspices a group from LACTOZ produced an extensive and useful review of the chemical scheme used in the EMEP photo-oxidant model (Wirtz *et al.*, 1994).

An intercomparison study has been performed on gas-phase mechanisms used in photochemical dispersion models inside and outside EUROTRAC (Poppe *et al.*, 1996)

b. Cloud models

Clouds comprise a multi-phase atmospheric system in which chemical and physical transformations of trace substances occur, and because they are ubiquitous in the troposphere they have to be treated adequately in simulation models. Clouds are complex physical and chemical systems, which are not well understood. Considerable experimental effort has been given to them in the subproject GCE and in an earlier subproject ACE which was devoted to elevated clouds. As well as these field experiments, chemical aspects of clouds have been studied in the laboratory in the subproject HALIPP, see section 5.3.

There are many models of cloud behaviour: in a recent overview some 21 cloud, fog and scavenging models of differing complexity are described. Some of them are directed to specific aspects of clouds, such as the life cycle of radiation fogs; others consider only one cloud type, for example convective clouds (Floßmann and Cvitaš, 1995). Of these 21 cloud models, three of the more general relatively simple ones are actually incorporated in the continental and global simulation models which have been discussed in the previous sections. For further details, see Chapter 3 and Table 3.2.

c. Photolysis processes

There have been several attempts to use field measurements to parameterise photolysis rates and to study the effects of clouds. At the University of Munich, a one dimensional fog model has been developed which has been incorporated into EURAD (Hass and Ruggaber, 1995).

5.4.7   Conclusions and recommendations for future applications

This chapter has described simulation models, developed, improved and applied in EUROTRAC, for both photo-oxidants and acidification or nutrification, and which may be applied to different scales, urban, continental and global.

Without doubt these simulation models are scientific tools which can be, and have been, used to study specific scientific questions. The same simulation models have been applied, primarily to investigate the relationship between precursor emissions, photo-oxidants and acidification or nutrification. They form one of the vital elements of EUROTRAC, in which new knowledge and experience can be integrated; they will also be used to answer policy-relevant questions.

An issue that can be addressed with these models is the analysis of observed trends in ambient concentrations in Europe, the causes of which are often unclear. Trends over the last decades have been observed and can be modelled using the information available on emission changes.

Monitoring for extended periods, trend measurements, are also an essential element in the evaluation of the success or failure of abatement strategies. As with historical trends, the application of simulation models in connection with monitoring is essential to an evaluation. Simulation models can also be used to define the minimum number of stations and their geographical position needed to monitor air quality and give air quality warnings.

Simulation models can also be used to assess the accuracy of current emission data bases such as those established in GENEMIS.

Thus in a direct sense, simulation models can be used to establish the effectiveness of proposed abatement strategies.

5.5   Other tools

This section indicates a number of other products from EUROTRAC which lay the ground for future scientific and policy-oriented work in this field.

5.5.1   Data bases from field experiments

There are six EUROTRAC data bases resulting from field experiments, from ALPTRAC, TOR, TRACT, BIATEX, ASE and GCE (Hahn and Borrell, 1996).

In ALPTRAC, field measurements were performed to determine pollutant concentrations in snow. Ice-core measurements were also made as well as precipitation measurements at three alpine mountain sites. The ALPTRAC data are collected in a central ALPTRAC data catalogue and are available at the University of Heidelberg.

An extensive data base has been established for the TOR project. Concentration measurements, mainly hourly data, are available for O_3, NO, NO_2, CH_4 and CO together with radiative flux and meteorological parameters for 27 stations covering the whole of Europe. Ten stations also report speciated VOC measurements. This data base, which also has the character of a trends and monitoring network for a number of stations, is housed at RIVM in the Netherlands. There are close links with the monitoring network from EMEP.

Within the TRACT subproject a major field campaign was conducted in September 1992 over south-western Germany, eastern France and northern Switzerland. The aim was to study the movement of pollutants through the Rhine valley and over the surrounding mountains. Some ten aircraft and motor gliders, a number of ground stations specially set-up for this study, and parts of the various national monitoring networks were involved in the experiment. The TRACT data base, which contains 11 sub-data bases ranging from aircraft measurements to an emission inventory and is kept at the University of Karlsruhe, is a valuable tool for the investigation of transport in orographically complex terrain.

A series of three experiments were carried out in the TRANSALP sub-section of TRACT to assess the extent to which the air masses penetrate and pass through the alpine barrier at lower levels. Tracers were released at selected sites and collected at sampling points downwind of the release and, in the last exercise, with an aircraft. The data base, which will be useful to those interested in transport through complex terrain, is available from the Turin Geophysical Institute.

Within BIATEX two data bases were created: a generalised description of deposition over Europe for a number of essential compounds, and an emission data base of biogenically produced hydrocarbons and other compounds. These two data bases can be seen as the integration and compilation of the experimental work performed in the field in BIATEX. In ASE, elements are available for a generalised description of deposition of some compounds over the sea, and for the establishment of an emission data base of biogenically produced DMS over the sea. The BIATEX and ASE deposition data bases can be used directly as sub-models in simulation models; the BIATEX and ASE biogenic emission data base could be incorporated directly in the GENEMIS European emission data base. The BIATEX data base is available at ECN and the ASE data base at Risø.

The results of cloud measurements performed in the subproject GCE are available in a data base for three experiments, the Po valley campaign in 1989, the Kleiner Feldberg cloud experiment in 1990, and the Great Dun Fell experiment in 1993. The data base is at the University of Bologna.

5.5.2 Emission and land-use data bases

Data bases for emissions and land-use are being gathered together in the subproject GENEMIS. The GENEMIS emission data base contains SO_x, NO_x,

VOC, NH_3 and CO for 1990 based on LOTOS and CORINAIR annual emission inventories. An emission model to calculate hourly emissions for the species mentioned, both for anthropogenic, most notably power plants and road traffic, and biogenic emissions is available. The grid resolution is 80×80 km^2. It should be noted that a reliable emission data base is a prerequisite for the use of simulation models.

A reliable land-use data base is also available in GENEMIS; it is essential for the determination of biogenic emissions and for the description of dry deposition, see section 2.2.3.

5.5.3   Monitoring networks

As has been mentioned in section 5.5.1, two networks of measuring stations were established in EUROTRAC. The ALPTRAC network covers deposition in the Alps, the TOR network covers Europe. Both are science-oriented networks where, at specific stations, process-oriented studies are performed.

Because of their extended existence in time, both networks could contribute to the policy-relevant monitoring of pollutant concentrations or deposition. This is clearly shown by the co-operation between EMEP and the TOR network.

5.5.4   Hidden tools

There are a number of further items which could be considered as EUROTRAC tools. A good example is the new methodology which has been developed in BIATEX to measure dry deposition more reliably. The detailed description of such a methodology is also a utilisable product.

5.5.5   A network of scientists

EUROTRAC consists of 14 subprojects. These subprojects contain a total of about 250 contributions that have been accepted by the Scientific Steering Committee as a useful and essential contribution to EUROTRAC. On average therefore, a subproject contains about 20 contributions.

There is strong interaction within each subproject in the form of annual subproject workshops at which progress is revised and goals set. Thus a total of 14 "people networks" containing about 20 principal investigators (PI) exist, who know each other personally, and who have frequent interactions. It should be noted that many principal investigators have about five co-workers, and co-workers from different PIs will also have direct contact with each other. Also the 14 subproject coordinators meet annually, and often a coordinator in one subproject is also a PI in another, which increases the interactions still further.

Also there are two inter-subproject working groups, a Cloud group and a Chemical Mechanism working group, which have brought together participants from a number of subprojects.

These networks of scientists are themselves a tool, without which EUROTRAC could not have existed. They will continue to be a force for scientific progress in this field.

It should be stressed that this scientific infrastructure is to a large extent also a policy-oriented infrastructure. A considerable fraction of these 250 investigators perform research in their own country, funded by their respective Ministries of Environment, and so actually perform application studies. Their current knowledge and understanding is based on their frequent EUROTRAC contacts which have lead to a common understanding of specific phenomena. Without doubt this will lead to a coherence, based on the best available knowledge, in the advice offered to various European governments. Indeed one of the principal investigators in EUMAC was the Minister of Environment in his own country. The coherent advice and common understanding is of invaluable worth to the development of European environmental policy.

5.6 References

Ancellet, G., Pelon, J., Beekmann, M., Papayannis, A., Mégie, G., 1991, Ground based lidar studies of ozone exchanges between the stratosphere and troposphere, *J. Geophys. Res.* **96**, 22401–22421.

Atkinson, R., 1994, Gas-Phase Tropospheric Chemistry of Organic Compounds, *J. Phys. and Chem. Ref. Data* **9**, 1–216.

Ball, S.M., Hancock, G., Winterbottom, F., 1994, Quantum Yields of Excited Singlet Oxygen from Photolysis of Ozone, in: G. Angeletti and G. Restelli (eds), *Physico-Chemical Behaviour of Atmospheric Pollutants, Air Pollution Research Report EUR 15609*, **1** EN, EC Brussels, pp. 190–196.

Barnes, I., Becker, K.H., Zhu, T., 1993, Near UV absorption spectra and photolysis products of difunctional organic nitrates, *J. Atmos. Chem.* **17**, 353.

Becker, K.H., 1994, The atmospheric oxidation of aromatic hydrocarbons and its impact on photo-oxidant chemistry, in: P.M. Borrell, P. Borrell, T. Cvitaš, W. Seiler (eds), *Proc. EUROTRAC Symp. '94*, SPB Academic Publishing bv, The Hague, pp. 67–74.

Becker, K.H., 1995, Private communication.

Bierbach, A., Barnes, I., Becker K.H., Wiesen, E., 1994, Atmospheric Chemistry of Unsaturated Carbonyls: Butenedial, 4-Oxo-2-pentenal, 3-Hexene-2.5-dione, Maleic Anhydride, 3/-/-Furan-2-one, and 5-Methyl-3/-/-furan-2-one, *Environ. Sci. Technol.* **28**, 715.

Bongartz, A., George, Ch., Kames, J., Mirabel, Ph., Ponche, J.L.and Schurath, U., 1994, Experimental determination of mass accomodation coefficients using two different techniques, *J. Atmos Chem.* **18**, 149–169.

Bongartz, A., Schweighofer, S., Roose, Ch. and Schurath, U., 1995, The mass accomodation coefficient of ammonia on water, *J. Atmos Chem.* in press.

Borrego, C., Coutinho, M., and Barros, N., 1995, Intercomparison of two meso-meteorological models applied to the Lisbon region, *Meteorology and Atm. Phys.* **57**(1-4), 21–29.

Bösenberg, J., 1994, Atmospheric processes and ozone profiles, in: P.M. Borrell, P. Borrell, T. Cvitaš, W. Seiler (eds), *Proc. EUROTRAC Symp. '94*, SPB Academic Publishing bv, The Hague, pp. 99–104.

Bösenberg, J., 1995, *private communication*.

Bösenberg, J., Ancellet, G., Barbini, R., and Milton, M., 1994, *EUROTRAC Annual Report part 7: TESLAS*, EUROTRAC ISS, Garmisch-Partenkirchen.

Botha, C.F., Hahn, J., Pienaar, J.J., van Eldik, R., 1994, Kinetics and mechanism of the oxidation of sulfur (IV) by ozone in aqueous solution, *Atmos. Environ.* **28**, 3207–3212.

Brassington, D., 1995, Tunable diode laser absorption spectroscopy for the measurement of atmospheric species, Spectroscopy in environmental science, Clarke, R.J.H. and Hester R.E., *Adv. in Spectroscopy*, **24**, J. Wiley, New York, 84–147.

Brassington, D., Fischer, H., Klemp, D., Mac Cleod, H., Tacke, M., and Werle, P., 1994, *EUROTRAC Annual Report part 7: JETDLAG*, EUROTRAC ISS, Garmisch-Partenkirchen..

Brion, J., Chakir, A., Daumont, D., Malicet, J. Parisse, C., 1993, High-resolution laboratory absorption cross sections of ozone. Temperature effect, *Chem. Phys. Lett.*, **213**, 10.

Builtjes, P.J.H. and Reynolds S.D., 1983, Modelling the effects of emission controls in the Netherlands, *Environmental International* **9**, 573–580.

Builtjes, P.J.H., 1992, Long-term ozone simulation project - Summary Report, *TNO-Rep.* R 92/240.

Buxton, G.V., McGowan, S., Salmon, G.A., 1992, The free radical chain oxidation of S(IV) in aqueous solution, in: P.M. Borrell, P. Borrell, T. Cvitaš, W. Seiler (eds), *Proc. EUROTRAC Symp. '92*, SPB Academic Publishing bv, The Hague, pp. 599–604.

Carleer, M., Colin, R., Guilmot, J.M., Simon, P.C., and Vandaele, A.C., 1994, UV-visible absorption cross sections of relevant atmospheric trace species, in: P.M. Borrell, P. Borrell, T. Cvitaš, W. Seiler (eds), *Proc. EUROTRAC Symp. '94*, SPB Academic Publishing bv, The Hague.

Carnuth, W., Kempfer, U., Lotz, R. and Trickl, T., 1992, First year of continuous ozone measurements with the IFU Lidar, in: P.M. Borrell, P. Borrell, T. Cvitaš, W. Seiler (eds), *Proc. EUROTRAC Symp. '92*, SPB Academic Publishing bv, The Hague, pp. 225–226.

Chang, J.S., Brost R.A., Isaksen I.S.A., Madronich S., Middleton P., Stockwell W.R. and Walcek C.J., 1987, A three dimensional Eulerian acid deposition model: physical concepts and formulation, *J. Geophys. Res.* **92**, 14681–14700.

Colin, R., Carleer, M., Simon, P.C., Vandaele, A.C., Dufour, P.and Fayt, C., 1991, Atmospheric Transport Measurement by Fourier Transform DOAS, *EUROTRAC Annual report: part 7 TOPAS*, EUROTRAC ISS, Garmisch-Partenkirchen, pp. 14–16.

Crutzen, P.J. and Gidel L.T., 1983, A two-dimensional photochemical model of the atmosphere, *J. Geophys. Res.* **88**, 6641–6661.

Cvitaš, T. and Kley, D., 1994, *The TOR Network*, EUROTRAC ISS, Garmisch-Partenkirchen.

De Vries, H., 1994, Local trace gas measurements by laser photothermal deflection: physics meets physiology, Ph.D. thesis, University of Nijmegen.

Dentener, F.J. and Crutzen P.J., 1993, Reaction of N_2O_5 on tropospheric aerosols: impact on the global distributions of NO_x, O_3 and OH, *J. Geophys. Res.* **D4**, 7149–7163.

Ebel, A., Elbern, H. and Oberreuter, A., 1993, Stratosphere-troposphere air mass exchange and cross-tropopause fluxes of ozone, in: Thrane, E.V., *et al* (eds), *Coupling processes in the Lower and Middle Atmosphere*, Kluwer Academic Publishing, pp. 49–65.

Ebel, A., Haas, H., Jakobs, H.J., Laube, M., Memmesheimer, M., Oberreuter, A., Geiß, H. and Kuo, Y-H., 1991, Simulation of the ozone intrusion caused by a tropopause fold and cut-off low, *Atmos. Environ.* **25A**, 2131–2144.

Eberhard, J., Müller, C., Stocker, D.W., Kerr, J.A., 1995, Isomerisation of Alkoxy Radicals under Atmospheric Conditions, *Environ. Sci. Technol.* **29**, 232.

Elias H., Götz, U., Wannowius K.J., 1994, Kinetics and mechanism of the oxidation of sulfur (IV) by peroxyomonosulfuric acid anion, *Atmos. Environ.* **28**, 439–448.

Eliassen, A. and Saltbones J., 1975, Decay and transformation rates of SO_2, as estimated from emission data, trajectories and measured air concentrations. *Atmos. Environ.* **9**, 425–429.

Eliassen, A., Hov Ø., Isaksen I.S.A., Saltbones J. and Stordal F., 1982, A Lagrangian long-range transport model with atmospheric boundary layer chemistry, *J. App. Meteorology* **21**, 1645–1661.

Exner, M., Herrmann, H. Michel, J.W., and Zellner, R., in: P.M. Borrell, P. Borrell, T. Cvitaš, W. Seiler (eds), *Proc. EUROTRAC Symp. '92*, SPB Academic Publishing bv, The Hague.

Fiedler, F., Anfossi, D., Friedrich, R., Gassmannn, F., Girardi, F., Jensen, N.O., Jochum, A., Lightman, P., Mohnen, V.A., Neiniger, B., Rosset, R., Schaller, E., Stingerle, A., Wanner, W., 1994, *EUROTRAC Annual Report part 10: TRACT*, EUROTRAC ISS, Garmisch-Partenkirchen.

Floßmann, A. and Möller, D., 1995, *Cloud Models and Cloud Chemistry*, EUROTRAC ISS, Garmisch-Partenkirchen.

Flossmann, A., and Cvitas, T., 1993, Inventory of cloud, fog and/or scavenging models, *Clouds, Models and Mechanisms*, EUROTRAC ISS, Garmisch-Partenkirchen.

Fuzzi, S., Hansson, H-C., Jaeschke, W., 1994, *EUROTRAC Annual Report part 6: GCE*, EUROTRAC ISS, Garmisch-Partenkirchen.

Giovannoni, J.M., Mueller, F., Clappler, A. and Russel, A.G., 1993, Further development of a comprehensive air quality model to include acqueous phase chemistry, for calculating concentrations of ozone as well as nitrogen and sulfur containing pollutants in Europe, *EUROTRAC Annual Report, part 5: EUMAC*, EUROTRAC ISS, Garmisch-Partenkirchen.

Grgic, I., Hudnik, V., Bizjak, M., Levec, J., 1993, Aqueous S (IV) oxidation III. Catalytic effect of soot particles, *Atmos. Environ.* **27A**, 1409–1416.

Guesten, H., Heinrich, G., Schmidt, R.W.H. and Schurath, U., 1992, A novel ozone sensor for direct eddy flux measurements, *J. Atmos. Chem.* **14**, 73.

Guillard, C., Delprat, H., Hoang-Van, C., Pichat, P., 1993, Laboratory study of the rates and products of the phototransformation of naphtalene adsorbed on samples of titanium dioxide, ferric oxide, muscovite and fly ash, *J. Atmos. Chem.* **16**, 47–59.

Hahn, J. and P. Borrell (eds), 1996; *EUROTRAC Data Handbook*, EUROTRAC ISS, Garmisch-Partenkirchen, in preparation.

Hass, H. and Ruggaber, A., 1995, Comparison of two algorithms for calculating photolysis frequencies including the effects of clouds, **57** *Meteor. Atmos. Phys.* 97–100.

Hass, H., 1991, Description of the EURAD chemical transport model version 2 (CTM2), Mitteilung aus dem Institut für Geophysik und Meteorologie der Universität, Köln, No. 83.

Hayman, G.D., 1994, The oxidation of biogenic hydrocarbons, in: P.M. Borrell, P. Borrell, T. Cvitaš, W. Seiler (eds), *Proc. EUROTRAC Symp. '94*, SPB Academic Publishing bv, The Hague, pp. 75–82

Hesse, K., Schlemm, A., Wunderlich, C. and Schurath, U., 1994, Flow characterisitics of extremely thin jets: mass accommodation coefficient of ozone on water, in: P.M. Borrell, P. Borrell, T. Cvitaš, W. Seiler (eds), *Proc. EUROTRAC Symp. '94*, SPB Academic Publishing bv, The Hague, pp. 1085–1089.

Hongisto, M., 1992, A simulation model for the transport, transformation and deposition of oxidized nitrogen compounds in finland. Technical description of the model, Finnish Meteorol. Inst. Air Quality Publication no. 14, Helsinki, pp. 55.

Isaksen, I.S.A. and Hov Ø., 1987, Calculation of trends in the tropospheric concentrations of O_3, OH, CO, CH_4 and NO_x, *Tellus* **39B**, 271–285.

Jakobs, H.J., Feldmann H., Hass H., Memmesheimer M., 1995, The use of nested models for air pollution studies: an application of the EURAD Model to SANA episode, **34** *J. Appl. Meteor.* 1301–1319.

Jenkin, M.E., Hayman, G.D., 1994, The OH radical initiated oxidation of isoprene; in: K.H. Becker (ed), *Mechanism construction Tropospheric Oxidation, Air Pollution Research Report,* EC, Brussels, in press.

Junkermann, W., Platt, U., and Volz A., 1989, *J.Atmos.Chem.* **8** , 203–227.

Kames, J. and Schurath, U., 1992, Alkyl nitrates and bifunctional nitrates of atmospheric interest; Henry's law constants and their temperature dependencies, *J. Atmos. Chem.* **15**, 79–95.

Kames, J. and Schurath, U., 1995, Henry's law and hydrolysis rate constants for PANs using a homogeneous gas phase source, *J. Atmos. Chem.* in press.

Kanakidou, M., Dentener F.J. and Crutzen P.J., 1993, A global three-dimensional study of the fate of HCFCs and HFCs in the troposphere, *Proc. STEP/AFEANS workshop*, Dublin, Ireland, 113–129.

Kempfer, U., Carnuth, W., Lotz, R., Trickl, T., 1994, A wide-range ultraviolet lidar system for tropospheric ozone measurements: development and application, *Rev. Sci. Instrum.* **65**, 3145–3164.

Kirchner, W., Welter, F., Bongartz, A., Kames, J., Schweighofer, S. and Schurath, U., 1990, Trace gas exchange at the air/water interface: measurements of mass accomodation coefficients, *J. Atmos Chem.* **10**, 427.

Kley, D., Beck, J., Grennfelt, P., Hov, Ø, Isaksen, I.S.A., Penkett, S.A., 1994, *EUROTRAC Annual Report part 9: TOR,* EUROTRAC ISS, Garmisch-Partenkirchen.

Klug, W., Gömer D. and Wortmann B., 1989, Application of several interregional air pollution models for the simulation of acid deposition within the PHOXA program, *Technical PHOXA. Report* IV.

Lagrange, J., Pallares, C., Wenger, G., Lagrange, P., 1993, Electrolyte effects on aqueous atmospheric oxidation of sulfur dioxide by hydrogen peroxide, *Atmos. Environ.* **27A**, 129–137.

Langmann, B., 1995, Einbindung der regionalen troposphaerischem Chemie in die Hamberger Klimamodellumgebung: Modellrechnungen und Vergleich mit

Beobachtungsdaten. PhD-Thesis, Max-Planck Institut fuer Meteorologie, Hamburg, Germany, pp. 109.

Larsen, S.E., Baeyens, W., Belviso, S., Buat-Menard, P., Collin, J-L., Donard, O.F.X., Harrison, R.M., Leeuw, G. de, Liss, P.S., Rapsomanikis, S., Schulz, M., 1994, *EUROTRAC Annual Report part 2: ASE*, EUROTRAC ISS, Garmisch-Partenkirchen.

Le Bras, G., Becker, K.H., Cox, R.A., Lesclaux, R., Moortgat, G.K., Sidebottom, H.W., Zellner, R., 1993, *EUROTRAC Annual Report part 8: LACTOZ*, EUROTRAC ISS, Garmisch-Partenkirchen.

Le Bras, G., Becker, K.H., Cox, R.A., Lesclaux, R., Moortgat, G.K., Sidebottom, H.W., Zellner, R., 1994, *EUROTRAC Annual Report part 8: LACTOZ*, EUROTRAC ISS, Garmisch-Partenkirchen.

Lelieveld, J., Crutzen P.J., and Brühl C., 1993, Climate effects of atmospheric methane, *Chemosphere* **26**, nos. 1–4, 739–768.

Lightfoot, P.D., Cox, R.A., Crowley, J.N., Destriau, M., Hayman, G.D., Jenkin, M.E., Moortgat, G.K., Zabel, F., 1992, Organic Peroxy Radicals: Kinetics, Spectroscopy and Tropospheric Chemistry, *Atmos. Environ.* **26A**, 1805–1964.

Memmesheimer, M., Tippke, J., Ebel, A., Haas, H., Jakobs, H.J., and Laube, 1991,On the use of the EMEP emission inventories for European scale air pollution modelling with the EURAD model, in *Proc. EMEP Workshop on Photo-oxidant modelling for Long-Range Transport in Relation to Abatement Strategies,* Berlin, **307**, 324.

Mohnen, V.A., Slemr, F., Kanter, H-K., Fiedler, F., and Corsmeier, U., 1993, Quality assurance and quality control in TRACT, *EUROTRAC Newsletter* **12**, EUROTRAC ISS, Garmisch-Partenkirchen, pp. 18–23.

Moortgat, G.K., 1994, Radical Reactions of NO_3 with HO_2 and RO_2; in: G. Angeletti and G. Restelli (eds), *Physico-Chemical Behaviour of Atmospheric Pollutants, Air Pollution Research Report EUR 15609,* **1** EN, EC Brussels, pp. 66–76.

Moussiopoulos, N., 1993, Athenian photochemical smog: Intercomparison of simulations (APSIS), *Env. Software* **8**, 3–8.

Moussiopoulos, N., 1994, *The EUMAC Zooming Model, Model structure and applications*, EUROTRAC ISS, Garmisch-Partenkirchen.

Mowrer, J. and Lindskog, A., 1991, Automatic unattended sampling and analysis of background levels of C_2–C_5 hydrocarbons, *Atmos. Environ.* **25A**, 1971–1991.

Nester, K., Panitz, H.J., Fiedler, F., 1995, Comparison of the DRAIS and EURAD model simulations of air pollution in amesoscale area, **57** *Meteor. Atmos. Phys.* 135 – 158.

Neavyn, R., Sidebottom, H., Treacy, J., 1994, Reactions of hydroxy radicals with polyfunctional group oxygen-containing organic compounds, in: P.M. Borrell, P. Borrell, T. Cvitaš, W. Seiler (eds), *Proc. EUROTRAC Symp. '94*, SPB Academic Publishing bv, The Hague, pp. 105–109.

Oreskes, N., Schrader-Frechette K., Belitz H., 1994, Verification, validation and confirmation of numerical models in the earth sciences. *Science* **263**, 641–646.

Papayannis, A., Ancellet, G., Pelon, J., Mégie, G., 1991, Multiwavelength lidar for ozone measurements in the troposphere and lower stratosphere, *Appl. Optics* **29**, 467–476.

Peeters, J., Boullart, W., van Hoeymissen, J., 1994, Site-specific partial rate constants for OH addition to alkenes and dienes, in: P.M. Borrell, P. Borrell, T. Cvitaš, W. Seiler (eds), *Proc. EUROTRAC Symp. '94*, SPB Academic Publishing bv, The Hague, pp. 110–114.

Petry H., Elbern, H., Lippert, E., Meyer, R., 1994, Three-dimensionalmesoscale simulations of airplane exhaust impact in a flight corridor, in: U. Schumann and D. Wurzel (eds), Impact of emissions from Aircraft and Spacecraft upon the atmosphere, Procceddings Intern. Sci. Colloquium, Cologne, Germany, 1994, 329–335.

Platt, U., 1994, Differential Optical Absorption Spectroscopy (DOAS) in: M.W. Sigrist (ed), *Air Monitoring by Spectroscopic Techniques*, J. Wiley, New York, pp. 27–83.

Platt, U., 1994, Spectroscopic Measurements of Tropospheric Species, in: G. Angeletti and G. Restelli (eds), *Physico-Chemical Behaviour of Atmospheric Pollutants, Air Pollution Research Report EUR 15609*, 2 EN, EC Brussels, 664–673

Ponche, J.L., George, C., Mirabel, P., 1993, Mass transfer at the air/water interface: mass accomodation coefficients of SO_2, HNO_3, NO_2 and NH_3, *J. Atmos. Chem.* **16**, 1–21

Poppe, D., 1995, private communication.

Poppe, D., Y. Andersson-Skold, A. Baart, P.J.H. Builtjes, M. Das, F. Fiedler, Ø. Hov, F. Kirchner, M. Kuhn, P.A. Makar, J.B. Milford, M.G.M. Roemer, R. Ruhnke, D. Simpson, W.R. Stockwell, A. Strand, B. Vogel, H. Vogel, 1996, Gas-phase reactions in atmospheric chemistry and transport models: a model intercomparison, EUROTRAC ISS, Garmisch-Partenkirchen 1996.

Raes, F., van Dingenen R., Wilson J. and Saltelli A., 1993, Cloud condensation muclei from dimethyl sulfide in the natural marine boundary layer. In *DMS: Ocean Atmosphere and Climate*, Kluwer Acad. Publ. 311–322.

Schell, D., Georgii, H.W., Maser, R., Jaeschke, W., Arends, B.G., Kos, G.P.A., Winkler, P., Schneider, T., Berner, A., and Kruiz, C., Intercomparison of fog water samplere, *Tellus* **44B**, 612–631.

Schikowski, M., and 13 co-authors, 1994, Snow pit ampling intercomparison at Weißfluhjoch, Switzerland, in: P.M. Borrell, P. Borrell, T. Cvitaš, W. Seiler (eds), *Proc. EUROTRAC Symp. '94*, SPB Academic Publishing bv, The Hague, pp. 716–720.

Sedlak, D., Hoigné, J., 1993, The role of copper and oxalate in the redox cycling of iron in atmospheric waters, *Atmos. Environ.* **27A**, 2173–2185.

Simon, P.C., Colin, R., Galle, B., Plane, J., Platt, U. and Pommereau J-P., 1994, EUROTRAC *Annual Report part 7: TOPAS*, EUROTRAC ISS, Garmisch-Partenkirchen.

Slanina, J., 1995, Interaction between atmospheric chemistry and biology: an essential aspect of environmental research, *Life Chem. Reports* in press.

Slanina, J., Duyzer, J.H., Fowler, D., Helas, G., Hov, Ø, Meixner, F.X., Struwe, S., 1994, EUROTRAC *Annual Report part 4: BIATEX*, EUROTRAC ISS, Garmisch-Partenkirchen.

Slemr, J., Dietrich, J., Sheumann, B., Koemp, P., Kern, M., Junkermann, W., and Werle, P., 1994, Intercomparisons of Formaldehyde measuring techniques,in: P.M. Borrell, P. Borrell, T. Cvitaš, W. Seiler (eds), *Proc. EUROTRAC Symp. '94*, SPB Academic Publishing bv, The Hague.

Stern, R. and Builtjes P., 1989, Application of an Eulerian dispersion model to three photochemical episodes over North Western Europe within the PHOXA program, *Technical PHOXA, Report* II.

Stern, R. and Scherer B., 1989, Application of a complex acid deposition/photochemical oxidant model to an acid deposition episode over North Western Europe within the PHOXA program, *Technical PHOXA, Report* III.

Stern, R., 1993, Development and application of a 3-Dimensional photochemical dispersion model using different chemical mechanisms, PhD thesis, Free Univ. Berlin (in German).

Stockwell, W.R., Middleton, P., Chang, S., 1990, The second generation regional acid deposition model chemical mechanism for regional air quality modelling 95 *J. Geophys. Res.* 16343–16367.

Sunesson,.J.A., Apituley, A. and Swart, D.P.J., 1994, Differential absorption lidar system for routine monitoring of tropospheric ozone, *Applied Optics*, 7045–7058.

Tabor, K., Gutzwiller, L., Rossi, M.J., 1993, The interaction of NO2 with amorphous carbon: Heterogeneous chemical kinetics, in: J. Peeters (ed), *Chemical Mechanisms Describing Tropospheric Processes, CEC Air Pollution Research Report* **45**, Guyot, Brussels, pp. 35–40.

Torres, L., 1991, Continuous measurement of isoprene emission rates and fluxes, EUROTRAC *Annual Report part 4: BIATEX*, EUROTRAC ISS, Garmisch-Partenkirchen.

Trickl, T., 1994, Ozone measurements: 1993 results and the future of the IFU Lidar, in: P.M. Borrell, P. Borrell, T. Cvitaš, W. Seiler (eds), *Proc. EUROTRAC Symp. '92*, SPB Academic Publishing bv, The Hague, pp. 341–343.

Vinckier, C., Compernolle, F., Van Hoof, N., Ashty, S., 1993, Role of α-pinene in the formation of greenhouse gases in the atmosphere, in: *Proc. Belgian Global Change Symposium*, Brussels 1993, 173–188.

Volz-Thomas, A. and Borrell, P., 1993, Proposed "Perca" and "Roxbox" intercomparison campaign, *EUROTRAC Newsletter* **12**, EUROTRAC ISS, Garmisch-Partenkirchen, 30.

Volz-Thomas, A. , 1995, private communication.

Warneck, P., 1995, private communication.

Warneck, P., Mirabel, P., Salmon, G.A., Vinckier, C., Zetzsch, C., 1994, *EUROTRAC Annual Report part 6: HALIPP*, EUROTRAC ISS, Garmisch-Partenkirchen.

Warneck, P., Ziajka, J., 1995, Reaction mechanism of the iron (III)-catalysed autoxidation of bisulfite in aqueous solution; steady state description for benzene as radical scavenger, *Ber. Bunsenges. Phys. Chem.* **99**, 59–65.

Wayne, R.P., Barnes, I., Biggs, P., Burrows, J.P., Canosa-Mas, C.E., Hjorth, J., Le Bras, G., Moortgat, G.K., Perner, D., Poulet, G., Restelli, G., Sidebottom, H., 1991, The Nitrate Radical, Physics, Chemistry and the Atmosphere, *Atmos. Environ.* **25A**, 1.

Werle, P., and Slemr, F., 1995, private communication.

Wirtz, K., Roehl, C., Hayman, G.D., Jenkin, M.E., 1994, *LACTOZ re-evaluation of the EMEP MSC-W photo-oxidant model*, EUROTRAC ISS, Garmisch-Partenkirchen.

Wyers, G.P., Otjes, R.P., Slanina, J., 1993, A continuous-flow denuder for the measurement of ambient concentrations and surface-exchange fluxes of ammonia, *Atmos. Environ.* **27A**, 1085–2090.

Zabel, F., 1994, Mechanistic Studies of Ozone Reactions with Alkenes in: G. Angeletti and G. Restelli (eds), *Physico-Chemical Behaviour of Atmospheric Pollutants, Air Pollution Research Report EUR 15609*, Vol. **1** EN, EC Brussels, 197–206.

Zabel, F., 1995, Unimolecular decompostion of peroxinitrates, *Z. Physikalische Chemie* **188**, 119–142 .

Zang, V., van Eldik, R., 1993, The reaction of nitric oxide with sulfur(IV) oxides in the presence of iron(II) complexes in aqueous solution, *J. Chem. Soc. Dalton Trans.* 111–118.

Zellner, R., 1993, VOC oxidation under high NO_x conditions: Mechanisms and ozone formation potential, in: P.M. Borrell, P. Borrell, T. Cvitaš, W. Seiler (eds), *Proc. EUROTRAC Symp. '92*, SPB Acedemic Publishing bv, The Hague, pp. 339–345.

Zetzsch, C., Behnke, W., 1992, Heterogeneous photochemical sources of atomic Cl in the troposphere, *Ber. Bunsenges. Phys. Chem.* **96**, 488–493.

Zimmerman, P.H., 1987, 'Moguntia', a handy global tracer model, *Proc. 16th NATO-CCMS, ITM*, Lindau.

Appendix A
The Application Project Description

(The project description was approved by the IEC in July 1993
and formed the basis for the project)

Following the review of EUROTRAC in 1991/1992, the International Executive Committee (IEC) has decided to set up an Application Project (AP).

The aim of the AP is to assimilate the scientific results from EUROTRAC and present them in a condensed form, together with recommendations where appropriate, so that they are suitable for use by those responsible for environmental planning and management in Europe. The AP will thus contribute directly towards fulfilling one of the stated objectives of EUROTRAC, that is

"to improve the scientific basis for future political decisions on environmental management in the European countries".

1. Structure

a. The project will be guided by an Application Steering Group (ASG) which will be responsible to the IEC and Scientific Steering Committee (SSC) for the successful conclusion of the project. The ASG will consist of the Chairmen and Vice-Chairmen of the IEC, two members appointed by the SSC, the Director of the ISS and the Scientific Secretary.

b. The AP will consist of three working groups, each of which will address a particular theme of importance in the tropospheric environment.

c. One working group (photo-oxidants, see section 4 below) will consist of a convener and three other qualified scientists. The other two groups (acidification and tools) will consist of a convener and two other scientists.

d. The scientific secretary will be an ex-officio member of each working group. *e.g.* Each group will work together under the general direction of the convener, who will be responsible for the production of the report from that group.

e. The conveners together with the scientific secretary are responsible for producing the final report from the Application Project.

f. The conveners and scientists will be appointed by the IEC, on the recommendation of SSC, following nomination by the ASG.

g. The activities and work of the AP will be co-ordinated by the ISS, who will ensure the necessary contacts between the working groups and the ASG and be responsible for publishing the final report.

2.   Mode of operation and final report

a Each working group will be expected to review the work done in the appropriate subprojects, meet with the subproject coordinators and other people whom they think fit, including those in environmental agencies, and, within the timescale specified by the ASG, produce a report on the particular issue which they have been asked to address. The report should, after consultation with the subprojects concerned, be submitted by the ASG to the SSC for comment, and then to the IEC for final approval.

b. The individual reports should assess the problems involved in the specified theme, assimilate the relevant scientific results from EUROTRAC and present them in a condensed form, together with recommendations where appropriate, so that they are suitable for use by those concerned with environmental planning and management in Europe.

c. The result of the AP will be a single report which combines the reports of the separate working groups and presents an overview together with an executive summary.

3.   Proposed timetable of work

6/93   Approval of project & budget by IEC.

7/93   Appointment of participants

10/93   Provision of draft Layout for discussion by SSC & coordinator meetings in 11/93

1/94   Presentation by conveners at Joint Meeting of IEC, SSC and subproject coordinators

4/94   Presentation by conveners at EUROTRAC Symposium

9/94   Meeting of ASG with AP scientists

3/95   Draft AP report available for distribution to subproject coordinators

5/95   SSC approval of Report

6/95   IEC approval of Report

9/95   Publication of AP Report.

The ASG will meet as seems appropriate during this period.

4.   Themes to be addressed by the AP

a. Photo-oxidants in Europe;

- in the free troposphere; the variation in the distribution of the ozone column density could also be considered here;

- in rural atmospheres;

- in urban atmospheres.

The contribution of the work in EUROTRAC towards the improvement of atmospheric monitoring in Europe should also be considered.

b. Acidification of soil and water and the atmospheric contribution to nutrient inputs;

- the variation in type and concentration of acidification as exemplified by the content of SO_2, NO_x, NH_3 and organic acids in precipitation should be considered.

c. The contribution of EUROTRAC to the development of tools for the study of tropospheric pollution, in particular:

- tropospheric modelling,

- new or improved instrumentation,

- provision of laboratory data.

d. The following topics should be considered by both the photo-oxidant and acidification groups

- Emissions, both man-made and biogenic.

- Deposition;

    i) to the land surface;

    ii) to European coastal waters;

    iii) to ecologically sensitive areas.

- Any climatic effects.

Application Project Description: Annexe A

Some points which the IEC considers should be taken into account by the AP working groups

1. The recommendations should be addressed to decision makers in environmental agencies, government departments, large industries and municipalities. It is important to bear the intended recipients in mind when deciding on the approach and level used in the report.

2. The environmental questions (appendix B) developed by the IEC and SSC after consultation with the countries should provide a check list of the issues of current interest. The report could try to identify questions which cannot yet be answered with the required accuracy or level of reliability.

3. It should be borne in mind that, for each potential pollutant, a decision maker is faced with the following general questions.

   - How can the pollutant and its environmental impact best be assessed?

   - What is the relative importance of different pollutant sources?

   - What is the scientific evidence available on which to base the best strategy required to reduce the level of the pollutant?

   - What are the possible consequences for other pollution associated with alternative strategies or activities?

4. While EUROTRAC is not an "effects" programme, the relationship of results and recommendations to the accepted values of critical loads and target values for environmentally important contaminants would be useful to those concerned with abatement strategies.

5. While three separate themes have been designated, there is clearly overlap between them. The IEC hopes that the three groups will work closely and flexibly together to facilitate the work, eliminate unnecessary duplication and ensure that the final report appears as a coherent whole.

6. The working groups should consider using the report of the IPCC on climate change and that of the Swedish Environmental Agency on second generation abatement strategies as possible models for the final report.

Project description: Annexe B

Atmospheric environmental questions appropriate to EUROTRAC

A list is given of the environmental questions, pertaining to the troposphere, which are thought to be of current concern to the various countries. They are grouped according to the overall objectives of EUROTRAC. An indication is also given of the possible subprojects which may be able to contribute to answering the questions.

Preamble

Following the desire expressed at Symposium '90 that efforts should be made to focus the work of EUROTRAC towards policy issues, the SSC produced a first list of possible questions which might find an answer within the EUROTRAC framework. It was imagined that the questions could be used in two ways; firstly to present the practical relevance of the work being done in EUROTRAC to the participating governments; and secondly to provide some guidance to the subproject coordinators and principal investigators in the EUROTRAC subprojects on possible directions to steer their work as the project progresses.

The IEC subsequently endorsed this approach and requested its members to consult with the authorities in their own countries on the contents of the list. This brought further questions to light. The IEC edited and classified the whole list and requested the SSC to suggest which subprojects would be suitable to address the questions. A sub-group of the SSC has now considered the questions further and finally produced the list shown below.

It should be noted that the SSC realises that not all the questions fall within the terms of reference of EUROTRAC. However the questions do appear to encompass the major concerns over the troposphere and, of course, they may suggest scientific aims for future projects which will surely follow EUROTRAC.

The questions are classified according to four main areas of environmental importance:

a. the chemical formation and transport of photo-oxidants (*e.g.* ozone) in the troposphere on a regional and hemispheric scale;

b. the formation of acidity in the atmosphere, particularly processes involving the presence of aerosols and clouds;

c. the interaction of pollutants with the biosphere and the influence of the uptake and release of some constituents by the biota.

d. climate change and the troposphere.

A.   *Formation and transport of photo-oxidants*

General   *EUMAC*

1. Quantification of the concentration and spatial distribution of atmospherically important trace constituents in the troposphere over Europe and their temporal change due to anthropogenic activities.

Transport   *EUMAC, GLOMAC, TOR, TRACT*

2a. What is the origin of photochemical oxidants in rural areas and in conurbations (summer smog)?

 b. What are the contributions of domestic or imported VOC and NO_x to the formation of photo-oxidants in particular areas?

 c. To what extent do emissions from central Europe contribute to tropospheric pollutants 'downwind' of the centre?

 d. What is the role of the Arctic as a seasonal reservoir for precursors of tropospheric ozone and other photo-oxidants over Europe?

 e. What is the effect of the long distance transport of ozone precursors from outside Europe, say from North America, on the concentration of ozone and other photo-oxidants in the troposphere over Europe?

Photo-oxidant formation   *GCE, EUMAC, HALIPP, LACTOZ, TOR*

3a. What is the functional relationship between the formation of ozone and the occurrence of NO_x and hydrocarbons in urban and rural atmospheres?

 b. What is the photo-oxidant creation potential (POCP) of different VOCs? What is the importance of the timescale (days or months) for POCPs?

 c. What are the effects of clouds on the formation and removal of photo-oxidants?

Control   *LACTOZ, TOR, EUMAC*

4a. How much must NO_x and VOC emissions be reduced to meet ambient target levels for O_3 and other photo-oxidants?

 b. What is the most effective way to control peak levels of ozone and other photo-oxidants?

 c. What would be the most effective reduction strategies for urban and rural air pollution and how will reductions in urban emissions affect neighbouring rural atmospheres?

 d. Should particular compounds be targeted in controlling VOC emissions?

Sources   *ASE, GENEMIS, EUMAC, LACTOZ, TOR*

5. What is the influence of the increasing number of aircraft on the concentration and distribution of pollutants in the upper troposphere and how does this influence the chemistry of the lower troposphere?

6. What is the contribution of marine emissions to the formation of photo-oxidants over Europe?

Prediction   *EUMAC, TOR*

7. How can we predict the severity of episodes, involving the build up of pollutants in urban and rural regions, and the effectiveness of actions taken to ameliorate them?

8. Is it possible to develop efficient distribution models for European applications; *e.g.* for a) immediate smog forecasts; b) development of strategies for emissions; c) calculation of the extensive deposition of acids in the environment; d) calculation of pollutant distributions in the case of serious accidents?

B.   Acid formation

Sources   *ASE, ALPTRAC, BIATEX, GENEMIS, GCE, HALIPP, LACTOZ*

1. What is the contribution of NO_x chemistry to the formation of acidity?

2. Is the "missing" background deposition of acidity in Scandinavia oceanic in origin or due to transport of pollutants into the region?

3. What is the influence of biogenic marine trace constituents on acidity and photo-oxidants over Europe?

4. To what extent does the influx of Saharan dust neutralise the anthropogenically produced acidity in southern Europe?

Processes   *GCE, EUMAC, HALIPP*

5. What is the role of clouds in transport and transformation over Europe?

Transport & control   *ASE, BIATEX, EUMAC, GENEMIS, HALIPP*

6a. To what extent is it necessary to reduce acid component emission in Eastern Europe in order to obtain air quality that does not exceed critical loads?

 b. What is the pollutant loading at a given site and how does it depend on the temporal and spatial distribution of emissions and the chemical composition of the pollutants?

7. What will be the effects on European air quality of the expected substantial reductions in pollutant emissions in the eastern part of Europe?

C.   Eutrophication: (biosphere/atmosphere exchange)

Deposition to the sea and mountains

ALPTRAC, ASE, BIATEX, GENEMIS, EUMAC

1. What is the fraction of dry and wet deposition into the biosphere and how does it depend on the local and regional conditions, and factors such as emissions, distances from sources *etc.*?

2a. What is the contribution of airborne deposition to the pollution in the North, Baltic and Mediterranean seas?

b. What is the airborne contribution to the concentration of nutrients in coastal waters?

c. What is the amount of air-sea exchange of persistent synthetic chemicals?

3. What are the main source regions for atmospheric pollutants in Alpine regions and the Uludag Mountains?

Biogenic emissions   *BIATEX, GENEMIS*

4a. Determination of the temporal and spatial distributions of emissions of environmentally related biogenic trace constituents *e.g.* hydrocarbons, NH_3 *etc.*

b. What is the relative importance of natural and man-made emissions of these constituents and how do they contribute to the acidity of clouds and photo-oxidant concentrations?

c. How do changes in land-use affect the release and deposition of trace constituents?

Processes   *BIATEX, GCE, HALIPP*

5. What are the relative roles of clouds and fog, and of rain in the 'occult' and 'direct' deposition of acidity on vegetation?

D.   Climate change and the troposphere

EUMAC, GCE, GLOMAC, TOR

1. What changes can be expected in the radiation balance of the atmosphere as the pollutant composition of the troposphere changes?

EUMAC, LACTOZ, TOR

2. What is likely to be the effect of potential climate change on the concentrations and on the spatial and temporal distributions of pollutants both within Europe and from outside Europe?

Appendix B

The Aims and Objectives
of EUROTRAC & its Subprojects

EUROTRAC: European Experiment on the Transport and Transformation of Environmentally Relevant Trace Constituents in the Troposphere over Europe.

Goal

* The main goal of EUROTRAC is to provide the basic information to understand and predict the distribution, chemical transformation and transport of pollutants in the troposphere over Europe.

Aims

* To increase the basic knowledge in atmospheric science.

* To improve the scientific basis for taking future political decisions on environmental management in the European countries.

* To promote the technological development of sensitive, specific and fast response instruments for environmental research and monitoring.

Scientific objectives

To study

* the chemistry and transport of photo-oxidants (*e.g.* ozone) in the troposphere and its perturbation by anthropogenic activities in Europe;

* the processes leading to the formation of acidity in the atmosphere, particularly those involving aerosols and clouds;

* the uptake and release of atmospheric trace substances by the biosphere, relevant to the formation of photo-oxidants and acidity.

Field Measurements

ALPTRAC (High Alpine Aerosol and Snow Chemistry Study)

Coordinator: Dietmar Wagenbach (University of Heidelberg)
previously Hans Puxbaum (Technical University, Vienna)

Scientific objectives: to understand

* the main physical and chemical processes responsible for the occurrence
  and accumulation of acidic and aerosol components in the high alpine
  region;

* the contribution of various source regions to the deposition of trace
  components and their geographical and seasonal trends.

GCE (Ground-based Cloud Experiments)

Coordinator: Sandro Fuzzi (CNR, Bologna)
previously John Ogren (NOAA, Boulder)

Scientific objectives:

* To determine the factors controlling acid-rain formation and oxidant and
  catalyst concentrations in cloud droplets.

* To identify the dominant European cloud types.

* To determine the importance of aerosols in cloud composition and
  nucleation.

TOR (Tropospheric Ozone Research)

Coordinator: Dieter Kley (KFA, Jülich)

Scientific objectives:

* To ascertain how much the mean ozone concentration over Europe is
  greater than that over northern mid-latitudes generally.

* To determine and to model the trends in ozone concentrations.

* To try to measure any transfer of ozone between the boundary layer and the
  free troposphere and between the troposphere and the stratosphere.

TRACT (Transport of Pollutants over Complex Terrain)

Coordinator: Franz Fiedler (KFK, Karlsruhe)

Scientific objectives:

* To study orographic effects on atmospheric transport and exchange processes within the boundary layer over complex terrain with special reference to turbulent dispersal, channelling and mountain induced wind systems.

* To estimate the handover of air pollutants from the atmospheric boundary layer to the free troposphere.

Laboratory Studies

HALIPP (Heterogeneous and Liquid Phase Processes)

Coordinator: Peter Warneck (MPI, Mainz)

Scientific objectives:

* To understand the processes which generate and control oxidant and product concentrations in atmospheric droplets.

* To study the transfer of free radicals across gas/liquid interfaces.

* To estimate the contribution of aerosols to the chemistry of trace gases and radicals including the photochemical surface production of active species.

LACTOZ (Laboratory Studies of Chemistry Related to Tropospheric Ozone)

Coordinator: Georges Le Bras (CNRS, Orleans)
previously Karl-Heinz Becker (University of Wuppertal)
and Tony Cox (University of Cambridge)

The aim is to provide kinetic and mechanistic data for:

* The detailed modelling of simpler organic and NO chemistry which influences ozone formation in the free troposphere.

* The formulation of models describing the chemistry of more complex organic and NO_x compounds occurring in the polluted boundary layer, and in air parcels transporting ozone precursors to the free troposphere.

Model Development

EUMAC (European Modelling of Atmospheric Constituents)

Coordinator: Adolf Ebel (University of Cologne)

Scientific objectives:

*   To develop a three-dimensional Eulerian gridpoint model to investigate the interaction, on a European scale, of the relationships between the sources and receptors of natural and anthropogenic trace constituents.

*   To apply the model to problems of transformation and transport of trace constituents, particularly in interpreting results from other EUROTRAC subprojects.

GLOMAC (Global Modelling of Atmospheric Chemistry)

Coordinator: Henning Rodhe (University of Stockholm)

Scientific objectives:

*   To develop a three dimensional model of the global troposphere for simulating the transport and transformation of trace constituents.

*   To use the model to answer questions concerning the sources and sinks of tropospheric ozone and its precursors; the influence of anthropogenic processes on composition and climate; the long range transport of sulfur compounds and other acidifying substances.

GENEMIS (Generation of European Emission Data)

Coordinator: Rainer Friedrich (University of Stuttgart)

Scientific objectives:

*   The development of methods & models, using actual meteorological and socio-economic data, to transform available annual emission data from existing databases into hourly and seasonal values suitable for use in EUROTRAC models.

*   Improvement and extension of existing data bases.

*   Preparation of scenarios for future emissions.

Instrument Development

JETDLAG (Joint European Development of Tunable Diode Laser Absorption Spectroscopy for Measurement of Atmospheric Trace Gases)

Coordinator: David Brassington (Imperial College, London)
previously Franz Slemr (IFU, Garmisch-Partenkirchen)

Scientific objectives:

* To develop new TDLAS components and techniques and TDLAS instruments for environmental measurements.

* To determine high resolution infrared spectra of compounds of environmental interest.

TESLAS (Joint European Programme for the Tropospheric Environmental Studies by Laser Sounding)

Coordinator: Jens Bösenberg (MPI, Hamburg)
previously Jacques Pelon (CNRS, Paris)

Scientific objectives:

* To develop ultraviolet, visible and infrared lidars.

* To measure ozone in the presence of aerosols.

* To measure aerosols in the troposphere.

TOPAS (Tropospheric Optical Absorption Spectroscopy)

Coordinator: Paul Simon (IASB, Brussels)
previously Jean-Pierre Pommereau (CNRS, Paris)

Scientific objectives:

* To develop long path measuring equipment for visible and UV spectroscopy.

* To quantify the role of oxidised nitrogen compounds in photochemical cycles.

* To test photochemical theory by the measurement of the concentrations of OH radicals and other important tropospheric species.

Biosphere/Atmosphere Exchange

ASE (Air-Sea Exchange)

> Coordinator: Søren Larsen (Risø National Laboratory, Roskilde)
> previously Peter Liss (University of East Anglia, Norwich)

Scientific objectives:

* To assess processes and rates of emission of trace gases in European marine environments.

* To quantify the production and removal of marine particles and aerosols, and their interaction with atmospheric gases.

BIATEX (Biosphere-Atmosphere Exchange of Pollutants)

> Coordinator: Sjaak Slanina (ECN, Petten)

Scientific objectives:

* To study the mechanisms for the uptake and production of trace constituents in relevant European ecosystems.

* To provide regional fluxes of these trace constituents on seasonal and annual scales.

oOo

The following subproject was closed in 1991.

ACE (Acidity in Cloud Experiments)

> Coordinator: Tony Marsh (Imperial College)

The aims are to determine, as a function of season:

* Bulk conversion rates for $S(IV)$ to $S(VI)$ and for NO_x to nitrate ion, and also the effects of organic compounds on these rates.

* The contributions from mechanisms involving peroxide, ozone and transition metal catalysts to the conversion of $S(IV)$ to $S(VI)$.

* Oxidant concentrations and radiation intensities throughout clouds.

Subject Index

A

ACE 208
  project description 208
acid formation 205
acidic deposition 99
acidification 6, 20, 99
  contribution of deposited nitrogen 6
  importance of sulfur and nitrogen 21
  soil 21
aerosol particles 111
aerosol surface reactions 167
alkaline emissions
  effect on acidification 101
ALPTRAC 46, 147, 171, 204
  network of measuring stations 200
  project description 204
analytical methods
  dimethylsulfopropionate (DMSP) 155
  dimethylsulfoxide (DMSO) 155
Application project 30, 32, 195
  acknowledgements 35
  aim 1, 33
  environmental questions 201
  executive Summary 1
  formation 32
  funding 32
  mode of operation 35
  project description 195
  report organisation 35
  themes 2, 33
  tools 34
Application Steering Group, ASG 35
aqueous chemistry 165
ASE 102, 115, 117, 118, 129, 147, 155, 208
  project description 208
atmosphere
  recipient for pollutants 13
  transport medium for pollutants 13
atmospheric boundary layer, ABL 13, 132
atmospheric lifetimes 57
  ammonia, NH_3 15
  nitrogen dioxide, NO_2 15
  sulfur dioxide, SO_2 15
  VOCs 15
atmospheric processes 104

chemistry in clouds 107
field observations 110
gas-phase chemistry 106
laboratory studies 109
automated VOC measurements 154

B

best available technology, BAT 14
BIATEX 52, 91, 102, 116, 117, 118, 147, 208
  instrument development 154
  project description 208
  testing facilites 156
Brundtland Commission 26
budget of NOx
  controls 72

C

C_3–C_8 nitrates 59
catalytic oxidation 109
Chemical Mechanism Working Group, CMWG 170
  scientific advances 170
chemical mechanisms 183, 196
chlorine atom formation from aerosol
  surface reactions 169
climate 40
climate change 22, 70, 207
Climate Change Convention 30
climatology
  ozone 58
  photo-oxidants other than ozone 58
cloud chemistry 106, 120, 168
Cloud Group, CG 170, 171
cloud processes 7, 131
cloud-droplet composition 117
clouds
  convective 122
  orographic 126
  stratiform 122
CO as a tracer 61
coastal waters
  atmospheric inputs of nitrogen 117
commercial instrument development 154

H

HALIPP   55, 106, 107, 147, 159, 164,
   171, 205
  project description   205
  scientific advances   165
  Table of experimental techniques   175
HCFC
  use of Moguntia model   193
Heilbron experiment   189
HELCOM (Helsinki Commission)   28
Henry's law coefficients   169
Henry's law equilibrium   109, 111
heterogeneous chemistry   164, 182
heterogeneous droplet and aerosol
  chemistry   159
heterogeneous reactions   110
homogeneous gas-phase chemistry   159
hydrogen peroxide, H_2O_2 49, 168

I

IEC   36
IIASA   28
impact of air traffic   71
instrument development within
  EUROTRAC   8, 146
instrument subprojects   147
interactions
  land atmosphere exchange and air
    chemistry   134
  nitrogen and sulfur compounds   100,
    115, 117
intercomparison campaigns
  chemical analysis in ALPTRAC   156
  for DOAS instruments   149
  for TDLAS instruments   153
  instruments in BIATEX   155
  TROLIX   151
Intergovernmental Panel on Climate
  Change   see IPCC
IPCC   23, 30, 70, 80, 94, 119, 129, 193
ITE   125

J

JETDLAG   152, 207
  detection limits for TDLAS
    measurements   153
  project description   207

L

laboratory studies
  nature   159
  within EUROTRAC   159
laboratory techniques
  developments   171
LACTOZ   55, 88, 147, 159, 164, 170, 205
  project description   205
  scientific advances   162
  Table of experimental techniques   171
leaf area index   52
lidar
  see experimental techniques
Los Angeles smog   23
LRTAP Convention on Long-range
  Transboundary Air Pollution   14, 24,
  27, 32, 102, 119, 127
  signed protocols   25
Lurman, Carter, Coyner (LCC)
  mechanism   190

M

marine conventions   28
marine ecosystems
  input of nitrogen compounds   127
marine eutrophication   28
mass accommodation coefficients   107,
  109, 169
measurement sites
  American Samoa   45, 70
  Birkenes, Norway   46, 59, 63
  Cape Point, South Africa   45, 70
  Delft, Netherlands   45
  Germany   74
  Great Dun Fell, UK   110
  Hohenpeissenberg, Germany   46
  Izaña, Tenerife   72
  Jungfraujoch, Switzerland   94
  Kleiner Feldberg, Germany   110
  Kollumerwaard, the Netherlands   45
  Mace Head, Ireland   45, 59, 75
  Mauna Loa, Hawaii   45, 74
  Melbourne   46
  Ny Ålesund, Norway   59
  Po Valley, Italy   110
  Schauinsland, Germany   59, 61, 62, 79,
    81
  South Pole   45, 71
  Strath Vaich, UK   59
  Svanvik and Jergul, Norway   59

T

U

V

Printing: Saladruck, Berlin
Binding: Buchbinderei Lüderitz & Bauer, Berlin